WORLD BANK OPERATIONS EVALUATIO

The World Bank Forest Strategy
Striking the Right Balance

Uma Lele
Nalini Kumar
Syed Arif Husain
Aaron Zazueta
Lauren Kelly

2000
The World Bank
Washington, D.C.

www.worldbank.org/html/oed

First edition October 2000

Photo credits: Cover, National Geographic Image Collection; page 15, 17, 27, 39; Still Pictures.

ISBN 0-8213-4841-8

Library of Congress Cataloging-in-Publication Data

The world bank forest strategy : striking the right balance / Uma Lele . . . [et al].
p. cm. — (Operations evaluation studies)
Includes bibliographical references.
ISBN 0-8213-4841-8
1. Forests and forestry—Economic aspects. 2. Forest policy. 3. World Bank. I. Lele, Uma J. II. World Bank operations evaluation study.
SD393.W65 2000
333.75—dc21 00-043862

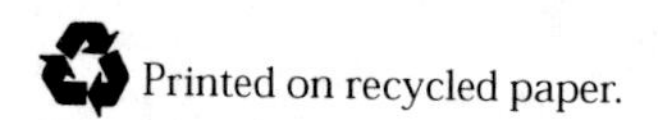

Contents

Figures

Study Team

Core Team
Uma Lele
Nalini Kumar
Syed Arif Husain
B. Essama Nssah
Aaron Zazueta
Lauren Kelly
Maisha Hyman

Additional Members
Madhur Gautam
Ridley Nelson

Consultants
Arnoldo Contreras-Hermosilla
Karin Perkins
Saeed Rana
Carolyn Barnes
Madelyn Blair
Kavita Gandhi

Brazil Country Team
Uma Lele
Virgilio M Viana
Adalberto Verissimo
Stephen Vosti
Karin Perkins
Syed Arif Husain

Cameroon Country Team
B. Essama-Nssah
James J. Gockowski

China Country Team
Scott Rozelle
Jikun Huang
Syed Arif Husain
Aaron Zazueta

Costa Rica Country Team
Ronnie de Camino Velozò
Olman Segura
Luis Guillermo Arias
Isaac Pérez

India Country Team
Nalini Kumar
Naresh Chandra Saxena
Yoginder K. Alagh
Kinsuk Mitra

Indonesia Country Team
Madhur Gautam
Uma Lele
Hariadi Kartodiharjo
Azis Khan, Ir. Erwinsyah
Saeed Rana

IFC Team
Afolabi Ojumu
Rafael Dominguez
Cherian Samuel
Dominique Zwinkels
John Gilliland

GEF Team
J. Gabriel Campbell
Alejandra Martin

MIGA Team
Harvey Van Veldhuizen

ENGLISH

FOREWORD

The changing dynamics of the forest sector and the global economy prompted World Bank President James Wolfensohn to launch the CEO Forum and the World Bank/ World Wide Fund for Nature Alliance. These initiatives offer the prospect of a proactive Bank role that would not have been conceivable a few years ago. In parallel, Bank management launched a Forest Policy and Implementation Review and Strategy process through the Environmentally and Socially Sustainable Development (ESSD) Network Vice Presidency. The Operations Evaluation Department (OED) was asked to contribute an independent evaluation of the Bank's 1991 Forest Strategy. Accordingly, OED has evaluated implementation of the strategy as outlined in *The Forest Sector: A World Bank Policy Paper* (1991), Operational Policy 4.36, and Good Practice 4.36 (both issued in 1993).

The evaluation reviewed lending and nonlending activities of the World Bank Group (International Bank for Reconstruction and Development [IBRD], International Development Association [IDA], International Finance Corporation [IFC], and Multilateral Investment Guarantee Agency [MIGA]) and the Global Environment Facility (GEF). The OED team carried out six country studies (Brazil, Cameroon, China, Costa Rica, India, and Indonesia), a global review, and six regional portfolio reviews (Sub-Saharan Africa, East Asia and the Pacific, Europe and Central Asia, Latin America and the

ESPAÑOL

PREFACIO

Los cambios en la dinámica del sector forestal y la economía global impulsaron al Presidente del Banco Mundial, James Wolfensohn, a iniciar el *CEO Forum* y la Alianza entre el Banco Mundial y el Fondo Mundial para la Naturaleza. Esas iniciativas permiten vislumbrar un papel proactivo para el Banco que hubiese sido inimaginable hace algunos años. Al mismo tiempo, la administración del Banco inició un proceso de examen de la ejecución de la política forestal y la estrategia para el sector, a través de la vicepresidencia de la Red sobre el Desarrollo Social y Ecológicamente Sostenible. Se solicitó al Departamento de Evaluación de Operaciones (OED) que aportara una evaluación independiente de la estrategia forestal del Banco para 1991. Por lo tanto, OED evaluó la aplicación de la estrategia conforme se reseña en *El sector forestal: un documento de políticas del Banco Mundial* (1991), Política Operacional 4.36, y Prácticas Optimas 4.36 (ambos publicados en 1993).

La evaluación examinó las actividades de financiamiento y no vinculadas con el financiamiento del Grupo del Banco Mundial (el Banco Internacional de Reconstrucción y Fomento [BIRF], la Asociación Internacional de Fomento [AIF], la Corporación Financiera Internacional [CFI], y el Organismo Multilateral de Garantía de Inversiones [OMGI] y el Fondo para el Medio Ambiente Mundial). El equipo del OED realizó seis estudios sobre países (Brasil, Camerún, China, Costa Rica, India e Indonesia), un examen mundial, y seis

FRANÇAIS

PRÉFACE

Les mutations en cours dans le secteur forestier et dans l'économie mondiale ont amené le président de la Banque mondiale, James Wolfensohn, à lancer le Forum des DG et l'Alliance Banque mondiale/Fonds mondial pour la nature. Ces initiatives permettent d'envisager pour la Banque un rôle dynamique qui n'aurait pas été concevable il y a quelques années. Parallèlement, la direction de la Banque a entrepris un « Bilan de la politique forestière et stratégie pour le secteur », par le biais du Réseau du développement écologiquement et socialement durable. Le département de l'évaluation des opérations (OED) s'est vu confier la tâche d'établir une évaluation indépendante de la stratégie forestière de la Banque de 1991. L'OED a donc évalué la mise en œuvre de la stratégie décrite dans *Le secteur forestier : Document de politique générale de la Banque mondiale (1991)*, la Politique opérationnelle 4.26 et la Pratique recommandée 4.36 (parues toutes deux en 1993).

L'OED a examiné les activités de prêt et les activités hors prêt du Groupe de la Banque mondiale (Banque internationale pour la reconstruction et le développement [BIRD], Association internationale de développement [IDA], Société financière internationale [SFI] et Agence multilatérale de garantie des investissements [AMGI]) et celles du Fonds pour l'environnement mondial (FEM). L'équipe de l'OED a effectué six études portant sur des pays (Brésil, Cameroun, Chine, Costa Rica, Inde et Indonésie), une analyse au niveau mondial et six analyses de portefeuilles

ENGLISH

Caribbean, the Middle East and North Africa, and South Asia). IFC, MIGA, and GEF findings are also incorporated in this report. The OED studies analyzed the *interactions* among the Bank's Country Assistance Strategies, economic and sector work, policy dialogue, and Bank lending. In addition to forest projects and projects with forest components, an attempt was made to assess the impact of adjustment lending operations and the indirect consequences of Bank operations in agriculture, environment, infrastructure, mining, transportation, electric power, energy, and oil and gas exploration. Over 700 operations were reviewed.

An extensive consultative process was carried out. Guidance was provided by an external Advisory Committee. OED, in collaboration with various in-country stakeholders, held workshops in Brazil, China, and India in November 1999 and in Indonesia in April 2000. The workshops were designed to allow governments and other country-level stakeholders (NGOs, the private sector, academia, and the like) an opportunity to comment on their country's case studies before the studies were offered to an international audience for feedback. OED placed the country background papers and the *Preliminary Report* on the Internet, after taking into account the comments of the Board's Committee on Development Effectiveness (CODE) at a meeting held on December 23, 1999. OED then held a Review Workshop in January 2000. Forestry experts, environmental activists, industry representatives, donors, and government policymakers met in Washington for two days to discuss the findings

ESPANOL

exámenes de la cartera regional (África al sur del Sahara, Asia oriental y el Pacífico, Europa y Asia central, América Latina y el Caribe, Oriente medio y Norte de África y Asia meridional). Se han incorporado a este informe, además, los resultados obtenidos por la CFI, el OMGI y el Fondo para el Medio Ambiente Mundial. OED analizó, en sus estudios, las *interacciones* entre las Estrategias del Banco de asistencia a los países, los estudios económicos y sectoriales, el diálogo de política, y el financiamiento del Banco. Además de los proyectos forestales y los proyectos con elementos forestales, se intentó evaluar el efecto de las operaciones de financiamiento para fines de ajuste y las consecuencias indirectas de las operaciones del Banco en la agricultura, el medio ambiente, la infraestructura, la minería, el transporte, la electricidad, la energía y la exploración de petróleo y gas. Se examinaron más de 700 operaciones.

Se realizó un amplio proceso de consulta. Un comité de asesoría externo tuvo a su cargo la orientación. OED, en colaboración con diversos interesados pertenecientes a cada país, realizó talleres en Brasil, China y la India en noviembre de 1999 y en Indonesia en abril de 2000. Los talleres se diseñaron para permitir que los gobiernos y los interesados a nivel del país (las ONG, el sector privado, el ámbito universitario, y otros) tuviesen la oportunidad de analizar los estudios de casos prácticos de su país antes de que esos estudios fuesen presentados para fines de información ante una audiencia internacional. OED publicó en Internet los documentos de antecedentes sobre los países y el *Informe preliminar*, después de tomar en cuenta los comentarios efectuados por el Comité del Directorio Ejecutivo

FRANCAIS

régionaux (Afrique subsaharienne, Asie de l'Est et Pacifique, Europe et Asie centrale, Amérique latine et Caraïbes, Moyen-Orient et Afrique du Nord et Asie du Sud). Les conclusions relatives à la SFI, l'AMGI et au FEM figurent aussi dans le rapport. L'OED s'est attaché, dans ses études, à analyser les *interactions* entre les Stratégies d'aide-pays, les analyses économiques et sectorielles, le dialogue de politique générale et les prêts de la Banque. En dehors des projets forestiers et des projets comprenant des composantes forestières, l'OED s'est efforcé d'évaluer également l'impact des opérations de prêt à l'ajustement et les conséquences indirectes des opérations concernant l'agriculture, l'environnement, les infrastructures, les activités extractives, les transports, l'électricité, l'énergie et l'exploration du pétrole et du gaz. Il a passé en revue plus de 700 opérations.

Ces travaux ont donné lieu à de vastes consultations. Un comité consultatif externe a suggéré les grandes orientations. En collaboration avec diverses parties prenantes nationales, l'OED a organisé des ateliers de travail au Brésil, en Chine et en Inde en novembre 1999 et en Indonésie en avril 2000, pour donner aux gouvernements et aux autres parties prenantes nationales (ONG, secteur privé, universités, etc.) la possibilité de présenter des observations sur les études de cas concernant leur pays avant qu'elles ne soient présentées à un auditoire international. L'OED a placé les documents de référence sur les pays et le rapport préliminaire sur l'Internet, après avoir pris en compte les observations formulées lors d'une réunion du Comité pour l'efficacité du développement (CODE) du Conseil, le

ENGLISH

of the OED review. Comments were also received through the Internet from a variety of stakeholders. The OED team also briefed President Wolfensohn. OED's *Preliminary Report*, country studies, and regional portfolio reviews were then discussed in the nine ESSD-organized regional and country-specific workshops in which OED participated. This final OED report was presented to CODE in June 2000 (The "Green Sheet" detailing CODE's comments is included in this volume as Annex M).

The main conclusion is that the Bank has implemented the 1991 Forest Strategy only partially, and mainly through an increased number of forest-related components in its environmental lending. The strategy sent a strong signal about changed objectives in the forest sector and included a new focus on conservation. However, its ambitious goals were not matched by commensurate means to implement the strategy. The controversy surrounding the policy formulation and implementation—including the ban on the use of Bank funds for all commercial logging in primary tropical moist forests—had a chilling effect on innovation. The effectiveness of the strategy has been modest, and the sustainability of its impact remains uncertain.

The country studies and the regional portfolio reviews contributed to OED's assessment of Bank operations through the lens of the 1991 Forest Strategy, as well as the perspectives of borrowers, Bank staff, the private sector, and some of the CEOs and nongovernmental organizations (NGOs) involved in a dialogue with the Bank. The ultimate purpose of the evaluation is to

ESPAÑOL

sobre la eficacia en términos de desarrollo, en una reunión celebrada el 23 de diciembre de 1999. Posteriormente, OED realizó un taller de examen en enero de 2000. Un grupo de especialistas en silvicultura, ecologistas, representantes de la industria, donantes y encargados de la formulación de políticas se reunió en Washington, D.C. durante dos días para analizar los resultados del examen del OED. Además, varios interesados hicieron llegar sus comentarios a través de Internet. El equipo del OED informó, además, al Presidente Wolfensohn. A continuación, el *Informe preliminar* del OED, los estudios sobre países y los exámenes de las carteras regionales fueron analizados en los nueve talleres regionales y específicos de cada país organizados por la Red sobre el Desarrollo Social y Ecológicamente Sostenible y en los cuales participó el mencionado departamento. El informe final del OED fue presentado al Comité sobre la eficacia en términos de desarrollo en junio de 2000 (El "Green Sheet" detallando los comentarios del Comité sobre la eficacia en términos de desarrollo se incluye en este volumen como Anexo M).

La conclusión principal es que el Banco ha aplicado la Estrategia para el sector forestal 1991 en forma parcial, y principalmente a través de una mayor cantidad de elementos vinculados con el sector forestal, en su financiamiento para el medio ambiente. La estrategia envió una clara señal acerca de los cambios en los objetivos del sector forestal e incluyó un nuevo enfoque de la conservación. Sin embargo, los medios para aplicar la estrategia no guardaban relación con sus metas ambiciosas. La controversia respecto de la formulación y la aplicación de la

FRANÇAIS

23 décembre 1999. L'OED a ensuite tenu un atelier de synthèse en janvier 2000, qui a donné à des experts forestiers, des écologistes, des représentants des milieux industriels, des bailleurs de fonds et des gouvernants, l'occasion de se réunir pendant deux jours à Washington pour débattre des conclusions de l'examen de l'OED. L'équipe de l'OED a également informé M. Wolfensohn du contenu de son rapport. Le rapport préliminaire de l'OED, les études par pays et les examens des portefeuilles régionaux ont ensuite été discutés au cours des neuf ateliers régionaux et nationaux organisés par le Réseau du développement écologiquement et socialement durable auxquels a participé l'OED. Le rapport final de l'OED a été présenté au CODE en juin 2000 (la « feuille verte » énumérant les observations du CODE figure dans l'Annexe M au rapport).

La principale conclusion est que la Banque n'a que partiellement appliqué la Stratégie forestière de 1991, et cela principalement dans le cadre de composantes forestières de prêts environnementaux. Elle a communiqué un message clair sur le recentrage des objectifs du secteur forestier en faveur de la préservation. Cependant, les moyens mis à disposition pour réaliser la stratégie n'étaient pas à la hauteur de ses ambitions. Les controverses qui ont entouré la formulation et la mise en œuvre de la stratégie, notamment l'interdiction d'utiliser des fonds de la Banque pour les activités d'abattage commercial dans les forêts tropicales humides primaires, ont découragé l'innovation. L'efficacité de la stratégie est demeurée limitée et la viabilité de son impact sur le long terme reste à démontrer.

ENGLISH

help shape the Bank's role regarding the forests of its borrowers at the dawn of a new millennium.

The report identifies seven elements that would make the Bank forest strategy more relevant to current circumstances and strengthen the Bank's ability to achieve its strategic objectives in the forest sector:

1. *The Bank needs to use its global reach to address both mechanisms and finances for international resource mobilization on concessional terms outside its normal lending activities.*
2. *The Bank needs to be proactive in establishing partnerships with all relevant stakeholders, governments, the private sector, and civil society to fulfill both its country and global roles.*
3. *The focus on primary tropical moist forests needs to be broadened to encompass all types of natural forests of national and global value.*
4. *Forest issues need to receive due consideration in all of the Bank's relevant sector activities and macroeconomic work.*
5. *Illegal logging needs to be reduced through the active promotion of improved governance and enforcement.*
6. *The livelihood and employment needs of all the poor need to be addressed, while continuing to safeguard the rights of indigenous people.*
7. *The Bank needs to align its organization and its resources with its strategic objectives in the forest sector.*

ESPAÑOL

política –que incluye la prohibición de utilizar los fondos del Banco para todo tipo de explotación forestal comercial en los bosques húmedos tropicales primarios– tuvo el efecto de desalentar la innovación. La estrategia ha tenido una eficacia modesta, y la sostenibilidad de su efecto es aún incierta.

Los estudios sobre países y los exámenes de la cartera regional contribuyeron a la evaluación de las operaciones del Banco efectuada por OED a través de la óptica de la Estrategia para el sector forestal 1991, como así también las perspectivas de los prestatarios, los funcionarios del Banco, el sector privado, y algunos de los CEO y las organizaciones no gubernamentales (ONG) que participaron en el diálogo con el Banco. El fin ulterior de la evaluación es ayudar a definir el papel del Banco respecto de los bosques de sus prestatarios al comienzo de un nuevo milenio.

El informe identifica siete elementos que harían que la estrategia del Banco para el sector forestal sea más acorde a las circunstancias actuales y que fortalecerían la capacidad del Banco para alcanzar sus objetivos estratégicos en el sector forestal:

1. *El Banco debe utilizar su campo de acción en el ámbito mundial para encarar tanto los mecanismos como las finanzas para la movilización de los recursos internacionales en condiciones concesionarias que no se encuentran dentro de la esfera de sus actividades normales de financiamiento.*
2. *El Banco debe ser proactivo en la formación de asociaciones con todos los interesados pertinentes,*

FRANÇAIS

Les études par pays et les analyses des portefeuilles régionaux ont permis à l'OED d'examiner les opérations de la Banque dans l'optique de la Stratégie forestière de 1991, ainsi que du point de vue des emprunteurs, du personnel de la Banque, du secteur privé et de certains des DG et organisations non-gouvernementales (ONG) qui ont participé au dialogue avec la Banque. Cette évaluation a pour but ultime de contribuer à déterminer le rôle qui revient à la Banque s'agissant des forêts des pays emprunteurs à l'aube du nouveau millénaire.

Le rapport identifie sept éléments qui permettraient de mieux adapter la stratégie forestière de la Banque aux circonstances actuelles et aideraient celle-ci à réaliser ses objectifs stratégiques dans le secteur forestier :

1. *La Banque doit mettre à profit ses moyens d'action mondiaux pour résoudre simultanément la question des mécanismes et celle des financements, s'agissant de la mobilisation de ressources internationales de caractère concessionnel en dehors de ses activités de prêt habituelles.*
2. *La Banque doit adopter une démarche dynamique pour établir des partenariats avec toutes les parties prenantes pertinentes (gouvernements, secteur privé et société civile) pour remplir son rôle aussi bien au niveau des pays qu'au niveau mondial.*
3. *La Banque doit élargir son action à tous les types de forêts naturelles ayant une valeur au plan national et mondial.*
4. *Il convient de tenir dûment compte des considérations forestières dans toutes les*

ESPAÑOL

los gobiernos, el sector privado, y la sociedad civil para cumplir su papel a nivel de cada país y a nivel mundial.

3. Es necesario ampliar el enfoque respecto de los bosques húmedos tropicales primarios con el objeto de abarcar a todos los tipos de bosques naturales que tengan un valor nacional o mundial.
4. Todas las actividades sectoriales y los estudios macroeconómicos pertinentes del Banco deben tomar en cuenta los temas forestales.
5. Se debe reducir la explotación forestal ilegal mediante la promoción activa de mejoras en la gestión de gobierno y la aplicación de las leyes.
6. Es necesario encarar las necesidades de sustento y empleo de todos los pobres, al mismo tiempo que se protegen los derechos de los pueblos indígenas.
7. El Banco debe adaptar su organización y sus recursos a los objetivos estratégicos del sector forestal.

FRANÇAIS

activités sectorielles et analyses macroéconomiques pertinentes de la Banque.

5. Il importe de réduire les coupes illégales en promouvant activement la gouvernance et le respect des lois et des règlements.
6. La Banque doit se pencher sur les besoins de tous les pauvres en matière de subsistance et d'emploi, tout en continuant à garantir les droits des populations autochtones.
7. La Banque doit harmoniser son organisation et ses ressources avec ses objectifs stratégiques dans le secteur forestier.

Robert Picciotto
Director-General, Operations Evaluation Department

Director-General, Operations Evaluation : *Robert Picciotto*
Director, Operations Evaluation Department : *Gregory Ingram*
Task Manager : *Uma Lele*

Acknowledgments

The preparation of this report would not have been possible without the support and valuable contributions of many individuals and organizations. The report benefited greatly from advice and intellectual guidance provided by an advisory committee of eminent experts: Conor Boyd, Angela Cropper, Hans Gregersen, and Emmy Hafild.

Colleagues both inside and outside the Bank challenged OED's work in progress, offered differing viewpoints and inputs, and made constructive suggestions. They include: Bagher Asadi, Mark Baird, Tulio Barbosa, Cornelis Baron van Tuyll van Serooskerken, Christopher Barr, Christopher Bennet, Eduardo Bertao, Hans Binswanger, Julian Blackwood, Juergen Blaser, Edward Bresnyan, Marjory-Anne Bromhead, Phillip Brylski, Bruce Cabarle, Mark Cackler, Gabriel Campbell, Jeff Campbell, Kerstin Canby, Wilfred Candler, Anne Casson, Gonzalo Castro, Kenneth Chomitz, Kevin Cleaver, Luis Constantino, Arnoldo Contreras-Hermosilla, Robert Crooks, Dennis de Tray, Peter Dewees, Mohan Dharia, Chris Diewold, James Douglas, Navroz Dubash, Hosny El-Lakany, Julia Falconer, Gershon Feder, Osvaldo Feinstein, Vincente Ferrer-Andreu, Douglas Forno, Rucha Ghate, Robert Goodland, George Greene, Jarle Harstad, John Hayward, Peter Hazell, Ian Hill, Gese Horskotte-Weseler, Korinna Horta, William Hyde, Peter Jipp, Ian Johnson, Norman Jones, Lisa Jordan, David Kaimowitz, Chris Keil, John Kelleberg, Irshad Khan, Robert Kirmse, Nalin Kishor, Odin Knudsen, John Lambert, Edwin Lim, Lennard Ljungmann, Tom Lovejoy, William Magrath, Dennis Mahar, Jagmohan Maini, William Mankin, Michael Martin, Alex McCalla, Glenn Morgan, Jessica Mott, Gobind Nankani, Ken Newcombe, Ilkka Juhani Niemi, Afolabi Ojumu, Keith Openshaw, Jan Cornelius Post, Idah Z. Pswarayi-Riddihough, V. Rajagopalan, Francisco Reifschneider, Emil Salim, Jeff Sayer, Ethel Sennhauser, Ismail Serageldin, Robert Schneider, Richard Scobey, Frances Seymour, Narendra Sharma, Susan Shen, Trayambkeshwar Sinha, John Spears, William Stevenson, Susan Stout, William Sunderlin, Wilfried Thalwitz, Guiseppe Topa, Hans Verolme, Jeurgen Voegle, Steve Vosti, Thomas Walton, Andrew White, Thomas Wiens.

Country authors served as sounding boards for the exchange of ideas, to solicit input from professionals and stakeholders, and to develop and test findings.

The team is grateful to colleagues in the Bank country offices who helped make the country workshops a success: John Garrison, Jin Liu, Neusa Queiroz, Elisa Romano, Ricardo L. B. Tarifa, Anis Wan, Melanie Widjaja, and Datin Yudha. Special thanks also go to the Brazilian Corporation for Agricultural Research (EMBRAPA) and the Ministry of Environment in Brazil; the Ministry of Environment and Forest in India; the Ministry of Finance, Badan Pengurusan Asrama, the government planning agency (BAPENAS), and the Ministry of Forestry in Indonesia; and the Ministry of Forestry, the State Development and Planning Commission, and the Ministry of Finance in China.

This report benefited considerably from the diverse perspectives of a truly broad set of commentators. The consultation process was facilitated by funding from the Swiss Agency for Development and Cooperation of the Royal Ministry of Foreign Affairs; the Department of Evaluation, Government of Norway; and the Ministry of Foreign Affairs, Government of Netherlands. OED also held a Forest Strategy Review Workshop in January 2000, involving 60 stakeholders from governments, the private sector, international and national nongovernmental organizations (NGOs), international and bilateral organizations, the professional community, and the Bank's senior management and staff. Detailed written and oral comments from the workshop have been taken into account in this volume. Participants are listed and highlights of their comments are presented in the Annexes. The comments in their entirety will be made available on the Internet.

Country workshops were held to discuss deaft country case studies in Brazil, China, India, and Indonesia. They involved particpation from stakeholder groups including government officials, NGOs, private sector, academics, and international agencies. In addition, OED participated in eight of the nine regional consultations organized by the Bank's Environmentally and Socially Sustainable Development Network (ESSD). The OED team's regular interaction with the ESSD forest team (Odin Knudsen, James Douglas, Juergen Blaser, Mariam Sherman, and Kerstin Canby) was useful in ensuring the review's client orientation.

Finally, the review team would like to acknowledge the contributions of Osvaldo Feinstein, Rema Balasundaram, Elizabeth Campbell-Pagé, Jacintha Wijesinghe, William Hurlbut, and Bruce Ross-Larson.

This publication was prepared by the Dissemination and Outreach team of the Partnerships and Knowledge Programs Group (OEDPK), under the guidance of Elizabeth Campbell-Pagé (task team leader), including Caroline McEuen (editor), Kathy Strauss (graphics and layout), and Juicy Qureishi-Huq (administrative assistant).

Uma Lele (Task Manager)
Nalini Kumar
Syed Arif Husain
Aaron Zazueta
Lauren Kelly

ENGLISH

EXECUTIVE SUMMARY

The Forest Sector: A World Bank Policy Paper (1991) presented a comprehensive statement of the Bank Group's forest strategy. Together with the associated operational policy, it brought the environmental agenda and participatory approaches to policymaking into the mainstream of the Bank's activities. It also challenged the Bank Group to adopt a multisectoral approach that would conserve tropical moist forests and expand forest cover.

But the strategy has been only partially implemented. Although it sent a strong signal about changed objectives in the forest sector and provided a new focus on conservation, the effectiveness of the strategy has been modest, and the sustainability of its impact is uncertain. Forest concerns have not been well integrated into Country Assistance Strategies, nor in the Bank's eco-

ESPANOL

RESUMEN

En *El sector forestal: un documento de políticas del Banco Mundial* (1991) presentó una declaración integral de la estrategia del Banco Mundial para el sector forestal. Ese documento, junto con la política operacional vinculada con el mismo, incorporó el programa para el medio ambiente y los enfoques participatorios de la formulación de políticas a las actividades principales del Banco. Además, instó al Grupo del Banco Mundial a adoptar un enfoque multisectorial tendiente a la conservación de los bosques húmedos tropicales y a la ampliación de la cobertura forestal.

Pero la estrategia fue aplicada en forma parcial únicamente. Aunque envió una clara señal acerca de los cambios en los objetivos del sector forestal e incluyó un nuevo enfoque de la conservación, la estrategia ha tenido una eficacia modesta, y la sostenibilidad de su efecto es aún

FRANCAIS

RÉSUMÉ ANALYTIQUE

Le secteur forestier : Document de politique générale de la Banque mondiale (1991) présentait de façon détaillée la stratégie forestière du Groupe de la Banque mondiale. Accompagné de la politique opérationnelle pertinente, il inscrivait le programme de travail environnemental et l'approche concertée de la définition des grandes orientations en la matière dans le cadre des activités courantes de la Banque. Il invitait en outre le Groupe de la Banque mondiale à adopter une démarche multisectorielle en vue de préserver les forêts tropicales humides et d'étendre le couvert forestier.

Or cette stratégie n'a été appliquée que partiellement. Elle a certes communiqué un message clair sur le recentrage des objectifs du secteur forestier en faveur de la préservation, mais son efficacité est demeurée limitée et la viabilité de son impact sur le long terme reste à démontrer. Les questions forestières ne sont pas bien intégrées aux Stratégies d'aide aux pays ni aux analyses économiques et sectorielles de la Banque. Enfin, la démarche multisectorielle envisagée n'a pas été suivie.

Les prêts accordés au secteur forestier ont augmenté de 78 %, mais ils représentent moins de 2 % des prêts de la Banque. Les opérations purement forestières, qui concernent traditionnellement des questions majeures liées à la politique et à la gestion forestières, plafonnent. La Banque vient d'incorporer des conditionalités forestières à quelques prêts à l'ajustement et elle a commencé

ENGLISH

nomic and sector work. The multisectoral approach envisaged has not been followed.

Although forest sector lending has increased by 78 percent, it remains less than 2 percent of overall Bank lending. Self-standing forest sector operations that traditionally deal with key forest policy and management issues have stagnated. The Bank recently introduced forest sector conditionality in a few adjustment loans and has begun to address issues of governance and corruption. But ensuring national ownership of reforms has been far more difficult. Much of the increase in lending has been in the form of forest components in agricultural or environmental projects. But these operations have typically been confined to the natural resource sector and have not addressed threats external to the forest sector that bring about forest and biodiversity loss. They have also increased the risk of the forest portfolio.

Forest-rich countries, the focus of the strategy, have sought to exploit their forests for legitimate development purposes, as well as for the benefit of powerful interest groups. As a result, the two central objectives of the strategy—slowing down rates of deforestation and increasing forest cover—have not been achieved. Some of the forest-poor countries, in contrast, have been ahead of the Bank in addressing problems of conservation and in incorporating forest concerns in overall development planning. In these countries, the Bank has helped in the realization of win-win outcomes. The countries have been able to alleviate poverty while improving, or at least minimizing, the loss of forest cover and biodiversity.

ESPAÑOL

incierta. Los temas forestales no se incorporaron en la debida forma a las Estrategias de asistencia a los países, y a los estudios económicos y sectoriales del Banco. No se siguió el enfoque multisectorial previsto.

Aunque el financiamiento para el sector forestal registró un aumento del 78%, aún representa el 2% del financiamiento total otorgado por el Banco. Las operaciones independientes del sector forestal que se ocupan, tradicionalmente, de los temas esenciales de la política y la gestión del sector forestal están estancadas. El Banco introdujo recientemente la condicionalidad del sector forestal en algunos préstamos para fines de ajuste y ha comenzando a encarar los temas de la gestión de gobierno y la corrupción. Con todo, ha sido mucho más difícil asegurar la identificación nacional con las reformas. La mayor parte del aumento en el financiamiento se produjo en forma de elementos forestales incluidos en los proyectos agrícolas o ambientales. Pero esas operaciones se limitaron, por lo general, al sector de recursos naturales y no se han ocupado de las amenazas externas al sector forestal que provocan pérdidas de bosques y de diversidad biológica. Además, han aumentado el riesgo de la cartera del sector forestal.

Los países con gran riqueza de bosques, que son el centro de la estrategia, han procurado explotar sus bosques con fines legítimos de desarrollo, como así también para beneficiar a grupos de intereses poderosos. En consecuencia, no se alcanzaron los dos objetivos principales de la estrategia: la disminución del ritmo de la deforestación y el aumento de la cobertura forestal. Por el contrario,

FRANÇAIS

à s'attaquer aux problèmes de la gouvernance et de la corruption. Mais il s'avère beaucoup plus difficile d'assurer l'adhésion nationale aux réformes. Une grande partie de l'augmentation des prêts est due à l'inclusion de composantes forestières dans des projets agricoles ou environnementaux. Cependant, ces opérations sont généralement limitées au secteur des ressources naturelles et ne concernent pas les menaces extérieures qui entraînent la perte de forêt et de diversité biologique. De plus, elles ont accru le risque associé au portefeuille forestier.

Les pays aux ressources forestières abondantes, sur lesquels est axée la stratégie, cherchent à exploiter leurs forêts à des fins légitimes de développement, mais aussi au profit de groupes d'intérêt puissants. Aussi, les deux objectifs principaux de la stratégie — ralentir le rythme du déboisement et étendre le couvert forestier — n'ont pas été atteints. Certains pays où les ressources forestières sont rares, en revanche, ont devancé la Banque dans ce domaine : ils se sont attaqués aux problèmes de préservation et ont inclus les questions forestières dans la planification générale du développement. Dans ces pays, la Banque a contribué à la réalisation d'objectifs bénéfiques pour tous. Les pays en question ont pu réduire la pauvreté tout en inversant, ou du moins en minimisant, la perte de couvert forestier et de biodiversité. En Afrique, cependant, où les pauvres sont le plus tributaires de la forêt, tant les activités économiques et sectorielles que les prêts au secteur forestier ont fortement diminué.

Il ressort de l'examen effectué par l'OED que ces carences et ces résultats s'expliquent par un certain nombre de

ENGLISH

At the same time, however, in Africa—where dependence of the poor on forests is the greatest—both forest sector economic and sector work and lending have declined sharply.

The OED review concluded that these implementation failures and outcomes are rooted in a number of limitations of the strategy and the associated operational policy.

First, the focus on tropical moist forests was too narrow. Other biodiversity-rich forests that are even more endangered, more important globally, and more critical to the survival of some 300 million forest-dependent people were neglected.

Second, while the strategy diagnosed the problem of externalities, it did not provide financing mechanisms to address the divergent costs and benefits of conservation at the local and global levels. At the local and national levels, communities and governments, given other pressing development imperatives and their limited ability to bear these costs, perceive the costs of conservation relative to their benefits to be higher than does the global community.

Third, the strategy failed to address governance issues, which have proved to be central—instead, confining itself to the narrow issues of economic incentive such as the length and price of concessions.

Fourth, the consultative process was too narrow. It overlooked the perspectives of important stakeholders such as governments, the private sector, and the civil society, which were essential in determining outcomes on the ground.

Fifth, the Bank lacked an internal implementation strategy and an incentive structure in terms of staff and

ESPAÑOL

algunos de los países que carecen de bosques han encarado los problemas de conservación antes que el Banco y han incorporado los temas forestales en la planificación general del desarrollo. En esos países, el Banco colaboró en el logro de resultados ventajosos. Los países pudieron reducir la pobreza y, al mismo tiempo, mejoraron o, por lo menos, minimizaron, las pérdidas de cobertura forestal y de diversidad biológica. Al mismo tiempo, sin embargo, en África –donde los pobres dependen en mayor medida de los bosques– tanto los estudios económicos y sectoriales como el financiamiento del sector forestal, han disminuido pronunciadamente.

En su examen, OED llegó a la conclusión de que los errores y las consecuencias de la aplicación están enraizados en varias limitaciones de la estrategia y de la política operacional vinculada con la misma.

En primer lugar, el enfoque en los bosques húmedos tropicales tenía poca amplitud. No se tomaron en cuenta otros bosques con gran riqueza de diversidad biológica que corren peligros aún mayores, que son más importantes a nivel mundial, y que son de vital importancia para alrededor de 300 millones de personas que dependen de los bosques para su subsistencia.

En segundo lugar, aunque la estrategia diagnosticó el problema de los efectos externos, no estableció mecanismos de financiamiento para encarar los costos y beneficios divergentes de la conservación en el ámbito local y mundial. A nivel local y nacional, las comunidades y los gobiernos, debido a otras necesidades acuciantes de desarrollo y a su capacidad limitada para soportar dichos costos, consideran, a diferencia

FRANCAIS

défauts de la stratégie et de la politique opérationnelle qui l'accompagne.

En premier lieu, la stratégie n'aurait pas dû se limiter aux forêts tropicales humides. Elle néglige d'autres forêts riches en biodiversité, qui sont encore plus menacées, plus importantes pour la planète et plus essentielles à la survie de quelque 300 millions de personnes qui en tirent leur subsistance.

En deuxième lieu, si la stratégie a bien diagnostiqué le problème des effets externes, elle ne prévoyait pas de mécanismes de financement permettant de s'attaquer aux divergences entre les coûts et avantages de la préservation au niveau local et au niveau mondial. Aux échelons local et national, compte tenu des autres impératifs de développement et du manque de ressources financières, les coûts de la préservation rapportés à ses avantages paraissent plus élevés aux collectivités et à l'État qu'à la communauté internationale.

En troisième lieu, la stratégie n'abordait pas les questions de gouvernance, dont il apparaît qu'elles ont joué un rôle essentiel, se bornant à traiter des aspects étroits des incitations économiques, tels que la durée et le prix des concessions.

En quatrième lieu, les consultations ont été trop restreintes. Elles n'ont pas permis de prendre en compte les positions de parties prenantes importantes telles que les gouvernements, le secteur privé et la société civile, positions qui expliquent en grande partie les résultats concrets obtenus sur le terrain.

En cinquième lieu, la Banque n'avait pas de stratégie d'exécution ni de régime d'incitations, au plan des ressources en personnel et au niveau de l'administration, pour faire face aux coûts de transaction élevés de la stratégie forestière. Au contraire, le

ENGLISH

administrative resources to support the high transaction costs of the strategy. Instead, the incentive structure in place worked against involvement in forest operations.

Sixth, there was insufficient foresight regarding the powerful forces of globalization and economic liberalization that are affecting forest outcomes.

And, finally, the Bank adopted an overly cautious approach in the wake of the controversy that surrounded the formulation and implementation of the policy. This discouraged risk-taking. Operational experience and the President's new initiatives already go beyond the Bank's 1991 strategy.

OED recommends that the Bank adopt a dual strategy:

- In its global role, the Bank can capitalize on its convening powers to facilitate partnerships that mobilize *additional* financial resources (over and above improved coordination of existing country-specific aid flows) for use in client countries, including new financing mechanisms of sufficient magnitude to achieve the global goals of the revised strategy.
- In its country-level role, the Bank can recognize and address the diverse realities in client countries, using all the instruments at its command and stressing long-term involvement, partnerships with a range of constituencies, learning by doing, and the exchange of experiences across countries. This entails a long-term commitment by the Bank, with enough resources for research, economic and sector work, and consulta-

ESPAÑOL

de la comunidad mundial, que los costos de la conservación son superiores a sus beneficios.

En tercer lugar, la estrategia no encaró los temas de la gestión de gobierno, cuya importancia ha quedado demostrada, y, por el contrario, se limitó a los temas puntuales de los incentivos económicos, como por ejemplo la duración y el precio de las concesiones.

En cuarto lugar, el proceso de consulta fue excesivamente limitado. No tomó en cuenta las perspectivas de los interesados importantes, como por ejemplo los gobiernos, el sector privado, y la sociedad civil, que eran esenciales para determinar las consecuencias sobre el terreno.

En quinto lugar, el Banco carecía de una estrategia interna de aplicación y de una estructura de incentivos para el personal, y de los recursos administrativos necesarios para soportar el alto costo de transacción de la estrategia. Por el contrario, la estructura de incentivos vigente desalentaba la participación en las operaciones forestales.

En sexto lugar, no se previó en forma suficiente el poder de las fuerzas de la globalización y de la liberalización económica que están afectando los resultados del sector forestal.

Y, finalmente, el Banco adoptó un enfoque extremadamente cauteloso cuando surgió la controversia alrededor de la formulación y la aplicación de la política, lo cual desalentó a las personas que estaban dispuestas a tomar riesgos. La experiencia operacional del Banco y las nuevas iniciativas del Presidente ya han superado la estrategia del Banco de 1991.

OED recomendó que el Banco adoptara una estrategia doble:

FRANCAIS

régime d'incitations en place était de nature à décourager les activités dans le secteur forestier.

En sixième lieu, la stratégie n'avait pas suffisamment prévu la montée des forces de la mondialisation et de la libéralisation économique, qui ont influé sur les résultats des opérations du secteur forestier.

Enfin, la Banque a fait montre d'une prudence excessive à la suite des controverses qui ont entouré la formulation et la mise en œuvre de la politique forestière, si bien que les services de la Banque ont évité de prendre des risques. L'expérience opérationnelle de la Banque et les nouvelles initiatives du président vont déjà plus loin que la stratégie de 1991.

L'OED recommande l'adoption d'une double stratégie :

- Au niveau mondial, la Banque peut user de son influence pour faciliter des partenariats, afin de mobiliser des ressources financières *additionnelles* (qui compléteront l'amélioration de la coordination des apports existants d'aide à des pays donnés) à l'intention des pays clients, en particulier à travers de nouveaux mécanismes de financement suffisamment bien dotés pour que soient atteints les objectifs internationaux de la stratégie révisée.
- Au niveau des pays, la Banque peut prendre en compte les particularités des pays clients, en recourant à tous les instruments dont elle dispose et en mettant l'accent sur une action à long terme, sur des partenariats avec diverses entités, sur l'apprentissage sur le tas et sur l'échange de données d'expérience

ENGLISH

tive processes complementary to, but independent of, its lending operations.

OED has identified seven elements that would make the revised Bank forest strategy more relevant to current circumstances and strengthen the Bank's ability to achieve its strategic objectives in the forest sector:

1. *Use the Bank's global reach to address both mechanisms and finances for international resource mobilization on concessional terms outside normal Bank lending activities.* Pursue measures such as the Prototype Carbon Fund and other concessional financing mechanisms to compensate countries that are producing forest-based international public goods such as biodiversity preservation and carbon sequestration.
2. *Establish partnerships with all relevant stakeholders to fulfill both country and global roles.* At the same time, the Bank must recognize the resource implications of meeting global objectives and using participatory approaches.
3. *Broaden the focus on primary tropical moist forests to encompass all types of natural forests,* including temperate and boreal forests and other highly endangered, biologically rich forests in the tropics: the *cerrados* and Atlantic forest of Brazil, tropical dry forests, and the Western Ghats of India. The revised strategy should recognize that natural forests alone need not serve all forest functions. Some important functions (including meet-

ESPAÑOL

- En su papel a nivel mundial, el Banco puede aprovechar su poder de convocatoria para facilitar las asociaciones que movilizan recursos financieros *adicionales* (además de mejorar la coordinación de los flujos de ayuda específica para cada país, ya existentes) para ser utilizados en los países clientes, y que incluyen nuevos mecanismos de financiamiento cuya magnitud es suficiente para alcanzar los objetivos mundiales de la estrategia revisada.
- En su papel a nivel de cada país, el Banco puede reconocer y encarar las diversas realidades de los países clientes, mediante el uso de todos los instrumentos a su alcance y el énfasis en la participación a largo plazo, en las asociaciones con una gran variedad de representados, aprendiendo mediante la práctica, y en el intercambio de experiencias entre los países. Esto entraña un compromiso a largo plazo por parte del Banco, con recursos suficientes para la investigación, los estudios económicos y sectoriales, y los procesos de consulta complementarios de sus operaciones de financiamiento pero independientes de las mismas.

OED identificó siete elementos que harían que la estrategia del Banco para el sector forestal sea más acorde a las circunstancias actuales y que fortalecerían la capacidad del Banco para alcanzar sus objetivos estratégicos en el sector forestal:

1. Debe utilizar el campo de acción del Banco a nivel mundial para encarar tanto los mecanismos

FRANÇAIS

entre pays. Pour cela, la Banque doit s'engager sur le long terme et consacrer des ressources suffisantes à la recherche, aux analyses économiques et sectorielles et aux processus consultatifs complétant de manière indépendante ses opérations de prêt.

L'OED a formulé sept recommandations en vue de mieux adapter la stratégie forestière révisée de la Banque aux circonstances actuelles et d'aider la Banque à réaliser ses objectifs stratégiques dans le secteur forestier :

1. Mettre à profit les moyens d'action mondiaux de la Banque pour résoudre simultanément la question des mécanismes et celle des financements, s'agissant de la mobilisation de ressources internationales de caractère concessionnel en dehors de ses activités de prêt habituelles. Rechercher des dispositifs tels que le Fonds prototype pour le carbone et d'autres mécanismes de financement concessionnels pour dédommager les pays qui produisent des biens publics internationaux fondés sur les forêts, tels que la préservation de la diversité biologique et la fixation du carbone.
2. *Établir des partenariats avec toutes les parties prenantes pertinentes pour remplir son rôle aussi bien au niveau des pays qu'au niveau mondial.* Parallèlement, la Banque doit tenir compte de ce qu'implique, sur le plan des ressources, la réalisation d'objectifs de caractère international selon des méthodes participatives.
3. *Élargir son action à tous les types de forêts naturelles,* y compris les

ENGLISH

ing export and urban demand, providing environmental services, and meeting the employment and livelihood needs of the poor) can be served by tree planting, and its expansion could also relieve pressure on natural forests.

4. *Give due consideration to forest issues in all relevant sector activities and macroeconomic work, and support activities that will help protect natural forests of national and global value.* The Bank should streamline efforts to promote forest conservation and development and align these efforts with the overall development goals and aspirations of its client countries. The synergy between development and conservation objectives needs to be recognized and actively promoted through tree planting on degraded forest and non-forest lands, energy substitution, end-user efficiency, research, technology, and dissemination.
5. *Reduce illegal logging by actively promoting improved governance and enforcement of laws and regulations.* This will require helping Bank borrowers to improve and implement existing laws and regulations. The mobilization of national stakeholders (especially civil society and the private sector) to demand, implement, and monitor improved governance practices will also be necessary.
6. *Address the livelihood and employment needs of* all *poor people, while continuing to safeguard the rights of indigenous people.* More attention needs to be given to the effects of the forest strategy on all the

ESPAÑOL

como las finanzas para la movilización de recursos internacionales en condiciones concesionarias que no se encuentran dentro de la esfera de sus actividades normales de financiamiento. Debe establecer medidas tales como el Fondo tipo para reducir las emisiones de carbono y otros mecanismos de financiamiento en condiciones concesionarias a fin de compensar a los países que producen bienes basados en los bosques, suministrados por el sector público a nivel mundial, como por ejemplo la preservación de la diversidad biológica y la fijación de carbono.

2. *Debe establecer asociaciones con todos los interesados pertinentes para cumplir su papel a nivel de cada país y a nivel mundial.* Al mismo tiempo, el Banco debe reconocer las consecuencias que tendrá el cumplimiento de los objetivos a nivel mundial y el uso de los enfoques participatorios, para los recursos.
3. *Debe ampliar el enfoque respecto de los bosques húmedos tropicales primarios con el objeto de abarcar a todos los tipos de bosques naturales,* inclusive a los bosques de zonas templadas y los bosques boreales y a otros bosques tropicales con gran riqueza biológica que se encuentran en grave peligro: los *cerrados* y el bosque atlántico de Brasil, los bosques secos tropicales, y los Ghats Occidentales de la India. La estrategia modificada debería reconocer que no es necesario que los bosques naturales sean los únicos que cumplen todas las funciones forestales. La forestación puede cumplir

FRANCAIS

forêts tempérées et boréales et aux autres forêts très menacées, riches en ressources biologiques : les *cerrados* et la forêt atlantique au Brésil, les forêts tropicales sèches et les Ghats occidentaux en Inde. Il convient d'affirmer, dans la stratégie révisée, que les forêts naturelles ne sauraient à elles seules remplir la totalité des fonctions forestières. Certaines d'entre elles (telles que satisfaire la demande à l'exportation et la demande urbaine, fournir des services environnementaux et répondre aux besoins des pauvres en matière d'emploi et de subsistance) peuvent être remplies par des plantations d'arbres, dont l'expansion peut également atténuer les pressions exercées sur les forêts naturelles.

4. Tenir dûment compte des considérations forestières dans toutes les activités sectorielles et analyses macroéconomiques pertinentes, et soutenir des activités qui contribueront à protéger les forêts naturelles ayant une valeur au plan national et mondial. Il conviendrait que la Banque rationalise les efforts menés en vue de promouvoir la protection et le développement des forêts et harmonise ces efforts avec les aspirations et les objectifs de développement généraux des pays clients. Il importe de reconnaître la synergie entre les objectifs de développement et ceux de la préservation, et de la promouvoir activement par la plantation d'arbres dans les forêts dégradées et sur les terres non forestières, par les énergies de substitution, par l'amélioration de l'efficacité au niveau des utilisateurs, par la

ENGLISH

poor, particularly to the conflicting needs of different user groups.

7. *Realign Bank resources with Bank objectives in the forest sector.* The Bank's internal incentives and skill mix need to be enhanced so that operational staff feel they have the support and confidence of Bank management and country borrowers and access to the human and financial resources needed to address the risky and controversial issues of the forest sector. The Bank must also diligently and routinely monitor compliance with all safeguard policies in its investment and adjustment lending.

ESPANOL

algunas funciones importantes (entre las que se incluye satisfacer la demanda de exportación y la demanda urbana, suministrar servicios ambientales, y satisfacer las necesidades de empleo y subsistencia de los pobres), y su expansión podría aliviar, además, la presión sobre los bosques naturales.

4. Debe tomar en cuenta los temas forestales en todas las actividades y estudios macroeconómicos pertinentes, y prestar apoyo a las actividades que ayuden a proteger los bosques naturales que son valiosos a nivel nacional y mundial. El Banco debería racionalizar los esfuerzos por promover la conservación y el desarrollo de los bosques y adaptar esos esfuerzos a los objetivos generales de desarrollo y a las aspiraciones de sus países clientes. Se debe reconocer y promover activamente la sinergia entre los objetivos de desarrollo y de conservación a través de la forestación de los bosques empobrecidos y las tierras sin bosques, la sustitución de energía, la eficacia de usuario final, la investigación, la tecnología y la difusión.
5. *Debe reducir la explotación forestal ilegal mediante la promoción activa de mejoras en la gestión de gobierno y la aplicación de las leyes y reglamentaciones.* Con ese fin, se deberá ayudar a los prestatarios del Banco a mejorar y aplicar las leyes y reglamentaciones vigentes. También será necesaria la movilización de los interesados nacionales (especialmente la sociedad civil y el sector privado) para exigir, aplicar y supervisar

FRANCAIS

recherche, la technologie et la diffusion.

5. *Réduire les coupes illégales en promouvant activement la gouvernance et le respect des lois et des règlements.* À cet effet, il faudra aider les emprunteurs de la Banque à améliorer et à mettre en œuvre les lois et les règlements existants. Il sera également nécessaire de mobiliser les parties prenantes nationales (tout particulièrement la société civile et le secteur privé) pour qu'ils exigent, appliquent et surveillent l'application de pratiques de gouvernance améliorées.
6. *Se pencher sur les besoins de* tous *les pauvres en matière de subsistance et d'emploi, tout en continuant à garantir les droits des populations autochtones.* Il convient d'accorder une attention accrue aux effets de la stratégie forestière sur tous les pauvres, et en particulier aux besoins contradictoires des différents groupes d'utilisateurs.
7. *Consacrer au secteur forestier des ressources à la mesure des objectifs.* Il importe d'améliorer les incitations internes et la gamme de compétences au sein du personnel, afin que les services opérationnels sachent qu'ils jouissent du soutien et de la confiance de la direction de la Banque et des pays emprunteurs, et qu'ils disposent des ressources humaines et financières dont ils ont besoin pour s'attaquer aux problèmes difficiles et épineux du secteur forestier. La Banque doit aussi s'assurer systématiquement et avec diligence de l'application de toutes ses politiques de protection dans ses prêts d'investissement et dans ses prêts d'ajustement.

ESPAÑOL

las prácticas de buena gestión de gobierno.

6. *Debe encarar las necesidades de sustento y empleo de todos los pobres, y al mismo tiempo proteger los derechos de los pueblos indígenas.* Se debe prestar más atención a los efectos de la estrategia del sector forestal en todos los pobres, en particular a las necesidades opuestas de los distintos grupos de usuarios.
7. Debe adaptar los recursos del Banco a sus objetivos en el sector forestal. *Se deben reforzar los incentivos internos y la combinación de especialidades del Banco para que el personal operativo sienta que cuenta con el apoyo y la confianza de la administración del Banco y de los países prestatarios y con el acceso a los recursos humanos y financieros que son necesarios para encarar los temas riesgosos y controvertidos del sector forestal. Además, el Banco debe supervisar en forma diligente y rutinaria el cumplimiento de todas las políticas de salvaguardia en su financiamiento para fines de inversión y de ajuste.*

GLOSSARY

Because this report evaluates the implementation of the World Bank's 1991 Forest Strategy and Policy, it retains most of the definitions used in the 1991 report *The Forest Sector: A World Bank Policy Paper* [*]. The sources of other definitions are: *Review of Implementation of Forest Sector Policy* (World Bank 1994a) [**]; *State of the World's Forests* (FAO 1999c) [†]; and *A Sustainable Forest Future?* (Pearce, Putz, and Vanclay 1999) [‡]; and *Sustaining Tropical Forests: Can We Do It, Is It Worth Doing?* Report of the Discussion Meeting held at Graves Mountain Lodge, Syria, VA, October 2–7, 1998 [§].

†**AFFORESTATION/REAFFORESTATION:** Establishment of a tree crop in an area from which it has always or long been absent.

***AGROFORESTRY:** Land use system in which woody perennials are used on the same land as agricultural crops or livestock in some form of spatial arrangement or temporal sequence.

BIOPROSPECTING: Identifying commercially or medically useful chemicals in living organisms.

****BOREAL FORESTS:** Forests located in areas with mean annual temperatures of less than -4°C, dominated by pine, fir, spruce, larch, and birch, and covering large areas of Canada, Russia, and Scandinavia.

***CARBON FIXATION; CARBON SEQUESTRATION:** The conversion by plants, through photosynthesis of atmospheric carbon dioxide, into organic compounds. Substantially changing forests by clearing, burning, and so on, increases the release of carbon-based gases into the atmosphere, thereby contributing to the greenhouse effect.

CERRADOS REGION: This plateau, covering 21 percent of the Brazil's land area, is the country's second-largest ecoregion after the Amazon. Its habitats include savanna, scrub, grasslands, and dry forest.

COMPREHENSIVE DEVELOPMENT FRAMEWORK (CDF): The CDF is a holistic approach to development adopted by the World Bank. It seeks a better balance in policymaking by highlighting the interdependence of all elements of development—social, structural, human, governance, environmental, economic, and financial. It emphasizes partnerships among governments, donors, civil society, the private sector, and other development actors. Perhaps most important, the country is in the lead, both "owning" and directing the development agenda, with the Bank and other partners each defining their support for the country's plans.

CLEAN DEVELOPMENT MECHANISM (CDM): Article 12 of the Kyoto Protocol calls for the establishment of the CDM to promote investment in sustainable energy projects. CDM is intended to encourage investment in sustainable energy projects in the developing world through investments in technology to reduce greenhouse gas emissions by an investor and a partner in a developing country. After certification requirements have been met, reductions in emissions would convert into "credits" for the environmental benefit produced by the investment, with the value of the credit shared by the investor and the partner.

***CLOSED FOREST:** Forest with a stand density greater than 20 percent of the area, and where tree crowns nearly contact one another.

***COMMON PROPERTY:** Tenure system whereby resources are collectively owned and managed and nonowners are excluded from access to the resource.

***CONSERVATION:** Rational and prudent management of natural resources to achieve the greatest benefit, while maintaining the potential of the resource to meet future needs.

§**CONSERVATION FORESTRY:** The application of verifiable good practices for the management of forest resources, including woodland and trees, in ways that are ecologically sound, economically viable, socially responsible, and environmentally acceptable and which do not reduce the potential of these resources to deliver multiple benefits over time. (See also *preservation forestry*)

‡**CONVENTIONAL LOGGING:** Conventional logging has come to be viewed as less concerned with forest regeneration through management—frequently lacking government control—and unsustainable, that is, not focused on long-term timber supplies.

***CONVERSION FOREST:** Forest assigned for conversion to agriculture or other non-forest use.

***COMMERCIAL LOGGING:** Extraction of timber in large quantities for industrial use or export markets.

COUNTRY ASSISTANCE STRATEGY (CAS): A broad development framework produced by the Bank in collaboration with the government and other stakeholders and tailored to individual country needs. The CAS is a central tool of the Bank's management and Board for guiding and reviewing the Bank Group's country programs and is an important benchmark for judging the impact of its work.

***DEFORESTATION:** Change of forest with depletion of tree crown cover to less than 10 percent. The clearing of forests and the conversion of land to non-forest uses.

***DEGRADATION:** Biological, chemical, and physical processes that result in loss of the productive potential of natural resources in areas that remain classified as forests. Degradation may be permanent, although some forests may recover naturally or with human assistance.

***DEPLETION:** Reduction in forest area or volume as a result of deforestation.

***DESERTIFICATION:** Degradation of the land that ultimately leads to desert-like conditions.

***DESIGNATED FOREST:** Forest legally set aside for preservation or production.

***ECOTOURISM:** Nature tourism.

‡ECONOMIC ASSESSMENT: Makes three potentially major adjustments to a financial analysis:

(1) The **first** modification adjusts *financial* costs and benefits to reflect *shadow prices.* A shadow price, for example, the price of labor or the exchange rate—differs from a financial price in that it reflects the true *opportunity cost* of the resources in question.
(2) The **second** modification adds in all environmental and social consequences that affect the well-being of anyone within the *nation.*
(3) The **third** modification constitutes a *global* analysis and would also include the gains and losses of people outside the country in which the forest is located.

ECONOMIC AND SECTOR WORK (ESW): ESW is analysis that underpins the Bank's lending operations, informs policy dialogue, responds to country requests for specific advisory tasks, and provides policy advice to the development community.

***EXTERNALITY:** A cost (or a benefit) of an economic activity by one party that is unintentionally imposed on (or received by) another party without compensation (or payment), and leads to inefficiencies in competitive markets.

***FARM FORESTRY:** People-oriented forestry that is carried out on private farmlands.

†FOREST: Ecosystem with a minimum of 10 percent crown cover of trees and/or bamboo, generally associated with wild flora and fauna and natural soil conditions and not subject to agricultural practices. Forests are in two categories:

- *Natural forests:* forests composed of tree species known to be indigenous to the area.
- *Plantation forests:* established artificially by afforestation on lands previously non-forested within living memory, or established artificially by reforestation on land that was forested, by replacement of the indigenous species with a new and essentially different species or genetic variety.

FOREST PROJECTS: Projects in the agriculture sector of the World Bank that are classified as forest projects.

FOREST-COMPONENT PROJECTS: Projects in various sectors of the World Bank (including environment) that have specific activities or components directly related to forests.

****JOINT FOREST MANAGEMENT (JFM):** The transfer of a share of benefits from government to rural communities in exchange for implementing agreed forest management programs in state forests.

‡**LOGGING:** The process of harvesting timber from a forest, logging has come to be used in the context of unsustainable cutting, which is cutting that is not focused on long-term timber supplies.

‡**MANAGEMENT:** Relates to the management of resources, inventorying, and yield calculation and to silvicultural practice (such as timber cutting).

***MARKET FAILURE:** A deviation from the conditions required for the efficient allocation of resources by a purely competitive market.

****NON-TIMBER FOREST PRODUCTS (NTFP):** Forest products, other than timber, such as fruits, medicines, nuts, and bushmeat.

***OPEN ACCESS:** Absence of ownership claims over resources, permitting and leading to uncontrolled and excessive attempts at appropriation and use.

***OPEN FOREST:** Forest in which the tree canopy layer is discontinuous but covers at least 10 percent of the area and in which the grass layer is continuous.

***PRESERVATION FOREST:** Forest designated for total protection of representative forest ecosystems in which all forms of extraction are prohibited.

***PRIMARY FOREST:** Relatively intact forest that has been essentially unmodified by human activity for the past 60 to 80 years.

‡**PRIVATE GAINS/LOSSES:** Refers to the private interests of the stakeholder; that is, what benefits him/her.

***PRODUCTION FOREST:** Forest designated for sustainable production of forest products.

***PROTECTION FOREST:** Forest designated for stabilization of mountain slopes, upland watersheds, fragile lands, reservoirs, and catchment areas. Controlled sustainable extraction of non-wood products could be allowed.

‡**REDUCED-IMPACT LOGGING:** Well-managed logging, usually supervised.

***REFORESTATION:** The replacement or establishment of a tree crop on forestland.

***SECONDARY FOREST:** Forest subject to a light cycle of shifting cultivation or to various intensities of logging, but that still contains indigenous trees and shrubs.

***SHIFTING CULTIVATION:** Farming systems in which land is periodically cleared, farmed, and then returned to fallow; synonymous with slash-and-burn or swidden agriculture.

****SOCIAL FORESTRY:** A term used to describe a type of project that was first developed in the late 1970s. Such projects included tree planting carried out as a community undertaking, and sometimes farm forestry as well, with a focus on production of fuelwood and poles. More recently, the term has been used to refer to any kind of forest activity in which poor people are the main beneficiaries.

‡**SOCIAL GAINS/LOSSES:** Refers to the wider social perspective; the jurisdiction may be local, national, regional, or global. In theory, local, national, and global perspectives on "social" gains/losses may diverge. National and global agencies should take the "social" standpoint, but it is well known that this is not always the case.

STAKEHOLDERS: Parties interested in and/or affected by an activity or policy.

***STUMPAGE OR ROYALTY:** Fee or price of standing trees before logging.

SUSTAINABLE FOREST MANAGEMENT (SFM): Several definitions are in use: *(i) The continuous flow of timber products or other specific goods or services, many of which may be essential for sustaining the livelihood of indigenous forest dwellers. *(ii) The continued existence of the current ecosystem. *(iii) The long-term viability of alternative uses that might replace the original ecosystem. (iv) Utilization of forests without undermining their use by present and future generations. Different systems of management are required for each category of forests, depending on the intended output. (v) A system of forest management that aims for sustained yields of multiple products from the forest over long periods. **(vi) Management of forests to achieve a continuous flow of forest products and services of all kinds.

‡SUSTAINABLE TIMBER MANAGEMENT: A forest management system that aims for sustained timber yields over long periods.

***SUSTAINED YIELD:** Production of forest products with an approximate annual balance between net growth and harvest.

****TEMPERATE FOREST:** Forest located in areas with mean annual temperatures between -3°C and 5°C, dominated by broad-leafed tree species. Temperate forests are characterized by heavy human intervention or conversion into plantations.

***TROPICAL DRY FOREST:** Open forest with continuous grass cover; distinguished from other tropical forests by distinct seasonality and low rainfall. Includes woody savannas and shrublands.

***TROPICAL MOIST FOREST:** Forest situated in areas receiving not less than 100 millimeters of rain in any month for two out of three years, with a mean annual temperature of 24°C or higher; mostly low-lying, generally closed.

WORLD BANK GROUP: For the purpose of this study, World Bank Group refers to the activities of the International Bank for Reconstruction and Development (IBRD), International Development Association (IDA), International Finance Corporation (IFC), and Multilateral Investment Guarantee Agency (MIGA).

ABBREVIATIONS AND ACRONYMS

AFR	–	Africa Region
AGR	–	Agriculture Department (World Bank)
APL	–	Adaptable Program Loan
ARPP	–	Annual Review of Portfolio Performance
ATO	–	African Timber Association
CAS	–	Country Assistance Strategy
CBO	–	Community-based organization
CDF	–	Comprehensive Development Framework
CDM	–	Clean Development Mechanism
CFI	–	Continuous forest inventory
CGE	–	Computable general equilibrium
CGIAR	–	Consultative Group on International Agricultural Research
CIFOR	–	Center for International Forestry Research
CODE	–	Committee on Development Effectiveness
DEC	–	Development Economics Department (World Bank)
DFID	–	Department for International Development
EAP	–	East Asia and Pacific Region
ECA	–	Europe and Central Asia Region
EIA	–	Environmental impact assessment
EMBRAPA	–	Brazilian Corporation for Agricultural Research (*Empresa Brasileira de Pesqisa Agropecuaria*)
ENV	–	Environment Department (World Bank)
ERR	–	Economic rate of return
ESSD	–	Environmentally and Socially Sustainable Development (Network)
ESW	–	Economic and sector work
FAO	–	Food and Agriculture Organization
FI	–	Financial intermediary
FRR	–	Financial rate of return
FSC	–	Forest Stewardship Council
GDP	–	Gross domestic product
GEF	–	Global Environment Facility
GNP	–	Gross national product
IADB	–	Inter-American Development Bank
IBRD	–	International Bank for Reconstruction and Development (World Bank)
ICR	–	Implementation Completion Report
ICRAP	–	International Center for Research in Agroforestry
ID	–	Institutional development
IDA	–	International Development Association
IFC	–	International Finance Corporation
IFF	–	Intergovernmental Forum on Forests
IFIA	–	Intermountain Forest Association
IFPRI	–	International Food Policy Research Institute
IITA	–	International Institute of Tropical Agriculture
IPF	–	Intergovernmental Panel on Forests
IRR	–	Internal rate of return
IUCN	–	International Union for the Conservation of Nature
LCR	–	Latin America and the Caribbean Region
LIL	–	Learning and Innovation Loan
M&E	–	Monitoring and evaluation

MNA	–	Middle East and North Africa Region
MIGA	–	Multilateral Investment Guarantee Agency
NFP	–	National Forest Program
NGO	–	Nongovernmental organization
NRM	–	Natural resource management
OCS	–	Operational Core Services (World Bank)
OED	–	Operations Evaluation Department
OEG	–	Operations Evaluation Group (IFC)
OP	–	Operational Policy
PA	–	Protected area
PAD	–	Project Appraisal Document
PAR	–	Performance Audit Report
PFE	–	Permanent forest estate
PPG-7	–	Rain Forest Trust/Pilot Program
PREM	–	Poverty Reduction and Economic Management (Network)
PSR	–	Project Status Report
QAG	–	Quality Assurance Group
RDV	–	Rural Development Department (World Bank)
RUTA	–	Regional Unit for Technical Assistance (Costa Rica)
SAR	–	South Asia Region
SDC	–	Swiss Agency for Development and Cooperation
SDV	–	Social Development Department (World Bank)
SFM	–	Sustainable forest management
SME	–	Small and medium-size enterprises
TFAP	–	Tropical Forest Action Plans
TMF	–	Tropical moist forest
UNDP	–	United Nations Development Program
WRI	–	World Resources Institute
WTO	–	World Trade Organization
WWF	–	World Wide Fund for Nature/World Wildlife Fund

All dollar ($) figures are in U.S. dollars.

The Challenges of Forest Strategy

Formulation of the World Bank's 1991 Forest Strategy was prompted by alarming estimates that deforestation was affecting 17 to 20 million hectares a year in the developing world and that tropical moist forests were shrinking inexorably.[1] There was also concern that the Bank's lending activities had contributed to these trends.[2] With environmental awareness growing, the Bank crafted and endorsed a conservation-oriented forest strategy. The strategy explored the complex relationships between such global concerns as biodiversity and climate change, and such national issues as soil and water conservation and the protection of indigenous peoples.

The strategy broke new ground in several respects (box 1.1). It reoriented Bank forest operations toward environmental sustainability. It was the first comprehensive sector strategy to bring the conservation agenda into the mainstream of World Bank Group activities. It emphasized the strong need to protect and conserve primary tropical moist forests, and identified 20 countries that were to be given particular attention in the Bank's country assistance because their tropical moist forests were threatened.[3] Eventually, the Operational Policy on forestry was classified as a safeguard policy (see Annex I).[4] This was also the first World Bank strategy formulated with the active participation of stakeholders outside the Bank. The consultative process focused largely on nongovernmental organizations—it did not secure a broad-based consensus among developing countries, the private sector, and civil society—but it opened the door for multistakeholder consultations in Bank policymaking.

The strategy assessed the importance of tree planting to meet the fuelwood and other basic needs of the poor. It also emphasized the importance of a multisectoral approach to forest issues, stressing that factors outside the forest sector may be more important in explaining deforestation than those within the sector. It proscribed Bank Group financing of commercial logging in primary tropical moist forests because of uncertainties about the valuation of forest environmental services, inadequate knowledge of sustainable forest management systems, and the irreversibilities associated with the loss of tropical moist forests. The financing of infrastructure projects (such as roads, dams, and mines) that might lead to loss of forests was made subject to rigorous environmental assessment. The ban on Bank financing of commercial logging and the independently developed requirement for environmental assessments of Bank projects sent a clear signal that the institution would not be involved in activities associated with deforestation in primary tropical moist forests.

Experience with the 1991 Forest Strategy

The intent of the 1991 Forest Strategy (box 1.1) is now generally reflected in the Bank's forest investments. The lessons of experience, and concurrent global trends, reinforce its aims. Although some key borrowers consulted during this OED review were unaware of

BOX 1.1. BANK FOREST STRATEGY: THE 1991 FOREST PAPER AND THE 1993 OPERATIONAL POLICY DIRECTIVE

The 99-page World Bank publication *The Forest Sector: A World Bank Policy Paper* was published in September 1991. This paper (henceforth the 1991 forest paper) represented the initial comprehensive statement of a new direction for the Bank's forest strategy. A two-page Operational Policy directive (OP 4.36, produced in 1993; see Annex I) reflected the policy content of the paper, and a Good Practices summary (GP 4.36; in Annex J) provided operational direction to Bank staff. The 1991 forest paper, the OP, and.the GP together are the subject of OED's evaluation.

In today's Bank terminology, the 1991 forest paper sets out a Bank strategy and the OP defines the policy, although some outsiders consider this distinction confusing at best. The 1991 forest paper gave guidance on policy directions, programmatic emphases, and good practices, and specified the principles and conditions for Bank involvement in the forest sectors of its client countries. As the first instance of significant outside stakeholder participation in the formulation of a Bank sector strategy, this document came to be viewed by the outside world and many Bank staff as embodying the new direction of the Bank's forest strategy. Both the Bank's Board and civil society referred to this document, as well as OP 4.36, when they asked OED for an independent evaluation of the Bank's forest policy.

The foreword for the 1991 forest paper was signed by then Bank President Barber Conable, but the Board was not asked to—and did not—comprehensively approve the paper. It did discuss the paper, however, and endorsed specific principles, including the ban on financing commercial logging in primary topical forests; the incorporation of forestry issues into the general policy dialogue and Country Assistance Strategy; and the promotion of international cooperation, policy and institutional reform, resource expansion, and forest preservation. The Board also endorsed the statement that "in tropical moist forests the Bank will adopt, and will encourage governments to adopt, a precautionary policy toward utilization . . . Specifically, the Bank Group will not under any circumstance finance commercial logging in primary tropical moist forests. Financing of infrastructural projects . . . that may lead to loss of tropical moist forests will be subject to rigorous environmental assessment as mandated by the Bank for projects that raise diverse and significant environmental and resettlement issues. A careful assessment of the social issues involved will also be required" (p. 19). Finally, the Board also approved a specific section on conditions for Bank involvement.

the Bank's 1991 Forest Strategy, the environmental movement in developing countries is supportive of its conservation aims and has become far more active.

Experience with the strategy indicates that some of its prescriptions were subject to many interpretations, so that results often differed from expectations. The effects of globalization and liberalization, the unexpected diversity and complexity of forest circumstances, and the large number of competing interests produced more variation in outcomes, and fewer universally applicable principles, than the framers of the strategy expected. The key to successful forest strategy has proved to be conflict resolution more often than simply avoiding harm or seeking to protect the vulnerable. Important errors of omission included the neglect of forests that are at least as endangered as the tropical moist forests—Brazil's Atlantic forests, forests in the Western Ghats of India, the biodiversity-rich temperate and boreal forests of Eastern Europe, and the tropical dry forests of Africa on which millions depend for their livelihoods. In the "new Bank," the central mission of poverty alleviation and sustainable development makes it imperative that the Bank's forest strategy be eclectic, with a focus balanced between conservation and development, consistent with its mission and the knowledge base currently at its disposal. The need for a less constraining, more proactive strategy is illustrated by the definition of primary forests adopted in the 1991 Forest Strategy: "relatively intact forest that has been essentially

Both the 1991 forest paper and the OP emphasize that the Bank will not finance commercial logging in primary tropical moist forests. The 1993 OP adds that the Bank "does not . . . finance the purchase of logging equipment for use in primary tropical moist forest" (para. 1a). The OP also makes a confusing statement that "in areas where retaining the natural forest cover and the associated soil, water, biodiversity, and carbon sequestration values is the object, the Bank may finance controlled sustained-yield forest management" (para. 1f). However, the 1991 paper stressed a lack of agreement on what constitutes sustainable forest management and offered three different definitions for it. All three include the management of forests for *multiple uses,* as distinct from timber production alone, to which logging normally refers (see Glossary). Although this provision in the OP to finance forest management under controlled sustained-yield conditions allows forest management under specific conditions (which the drafters of the OP thought gave the Bank some flexibility), the OP raised more questions than it answered. According to survey results, Bank staff do not consider the OP to be flexible on this point. (Outsiders who reviewed this report during the consultation process considered the shift from "policy" to "strategy" confusing at best.) The Bank will need a more focused forest strategy paper and a clear Operational Policy consistent with the strategy if its future lending and nonlending activities are to result in improved forest management practices. This report argues that current practices tend to be environmentally destructive and socially inequitable in many countries. What constitutes "sustainable" forest management remains unresolved and location-specific, although the debate on forest management has moved on, using some universally agreed-upon criteria and indicators.

Based on the 1991 strategy statement, the OP also states that "the Bank distinguishes investment projects that are exclusively environmentally protective . . . or supportive of small farmers . . . from all other forestry operations." It goes on to say that "projects in this limited group may be pursued only where broad sectoral reforms are in hand, or where remaining forest cover in the client country is so limited that preserving it in its entirety is the agreed course of action." This report recommends that the Bank could more usefully and proactively work with stakeholders sympathetic to reform in borrowing countries, rather than waiting for reforms to be put in place before becoming engaged in the sector.

unmodified by human activity for the past 60 to 80 years." But most primary forests in developing countries have some human activity. By using so restrictive a definition, the strategy implied the preservation of primary tropical moist forests, rather than their conservation. Extending this definition to boreal and temperate forests would pose problems for the Bank's conservation and production activities even in those forests.

The 1991 forest paper did identify a fundamental problem: national governments (as well as individuals and community groups) often want to realize the capital found in standing trees when a wider concern for the global environment dictates the conservation of forests and the protection of biodiversity.[5] But the strategy did not address the implications of these conflicting priorities or the implicit gap in global public resources. It mentioned only the Global Environment Facility, which has limited resources and no mandate to compensate countries for the potential loss of income from forest protection. The 1991 paper failed to recognize the scale of the public goods dilemma, which requires the global community to pay forest owners to preserve natural forests. A rational solution to payments for environmental services is required to ensure the conservation of natural habitats of international and national importance. Such payments would be connected to the negotiation of transparent agreements based on broadly understood rules that are both enforceable and enforced.

BOX 1.2. THE WORLD BANK'S FOREST STRATEGY AT A GLANCE

The goal of the Bank's forest sector strategy was to address the twin challenges of rapid deforestation, especially of tropical moist forests, and the inadequate planting of new trees to meet the rapidly growing demand for wood products. These challenges were viewed as being connected to five key factors:

- Externalities that prevented market forces from achieving socially desired outcomes
- Strong incentives, particularly for the poor, to cut trees
- Weak property rights in many forests and wooded areas
- High private discount rates among those encroaching on the forests
- Inappropriate government policies, particularly concession arrangements.

Five principles were proposed for Bank involvement in the forest sector:

- A multisectoral approach
- International cooperation
- Policy reform and institutional strengthening
- Resource expansion
- Land use controls (including zoning, demarcation, and controls associated with tenure issues) to preserve intact forests.

Bank-financed activities were expected to comply with seven conditions:

- No Bank Group financing for commercial logging in primary tropical moist forests
- The adoption of policies and an institutional framework consistent with sustainability
- A participatory approach to the management of natural forests
- The adoption of comprehensive and environmentally sound conservation and development plans, with the roles and rights of key stakeholders, including local people, clearly defined
- Basing commercial use of forests on adequate social, environmental, and economic assessments
- Making adequate provisions to maintain biodiversity and safeguard the interests of local people, including forest dwellers and indigenous peoples
- Establishing adequate enforcement mechanisms.

Uncertain valuations and weak cause-and-effect relationships make assessing the costs and benefits of forest conservation contentious. The benefits are difficult to measure, value, or capture; occur in the long run; and are global, national, and local. The costs, however, tend to be local and immediate.[6] The Bank's economic and sector work in Costa Rica, for example, found more than 60 percent of the benefits of conservation in that country to be global. Conservation activities are increasingly being stressed in developing countries and where such activities are of national importance. In China, Colombia, and Costa Rica, the governments—on their own initiative and within their limited means—have already begun to undertake investments in watersheds of national importance. However, the international community's willingness to pay for conservation activities in developing countries remains low (Pearce and others 1999, Annex 2)—and appears to be even lower than assumed in the 1991 Forest Strategy. The consultative process for this OED study found that many developed country nationals and some international nongovernmental organizations tend to consider all the benefits of conservation to be national or local and question the need for grant financing. Others, however, acknowledge that the Bank cannot achieve global objectives without such financing.

Deforestation and degradation of natural forests continued in many countries throughout the 1990s. Countries with tropical moist forests have continued to log their forests on a massive scale, often illegally and unsustainably, because of the higher returns to alternative land uses made possible by a synergy among agricultural technologies, trade liberalization, and infrastructure investments (see Brazil, Indonesia, and Cameroon country studies). The Bank's decision not to finance commercial logging in tropical moist forests does not appear to have made any difference in the extent of logging. In many countries there is as much (or more) illegal logging as legal production. In forests already being exploited in an environmentally and socially unsustainable manner, the Bank has therefore often lost the opportunity to improve forest management. Little progress can be made until the valuation of standing trees and the associated biodiversity, products, and services they provide—reflects the "real" value of forests to society, and until institutions of governance are strong enough to control illegal logging and to enforce forest management using well-defined criteria and indicators. In many forest-rich countries demand for timber, a major cause of deforestation, is far more attributable to the processes of industrialization, urbanization, and the demand for exports than to fuelwood consumption by the poor. The 1991 forest paper did not anticipate these domestic and international demand trends and assumed that developing countries would meet their urban needs through imports from temperate countries.

In addition to analytic and political challenges, implementation of the strategy depends on the incentives affecting various actors. Governments often derive revenues from forests because, in most developing countries, forests are owned by the state. In some cases, such as China and India, government-mandated arrangements for marketing and processing forest products work against improved forest management. When individuals connected with governments benefit financially from such arrangements, as in Indonesia, the outcomes can be at odds with broader socioeconomic and environmental goals. Decentralization of power from central or provincial governments, without payments for environmental services, leads to deforestation when local objectives conflict with national or international policy goals.

Policy and institutional design is therefore crucial for forest strategy implementation. If forest sector strategies are to be effective, they must suit the specific geographic, biophysical, demographic, sociocultural, and economic circumstances for which forest interventions are designed. The coordination and relative balance among considerations of forest management, poverty alleviation, indigenous rights, and economic development depend on agro-ecological circumstances, population densities, levels of development, and political will. Bank strategy should be flexible enough to adapt to diverse circumstances.

An important issue being debated globally is how to achieve a balance between the developmental and environmental roles of forests. The principles of sustainable forest management range from the most basic (focusing only on the continuous flow of forest products) to the more complex (including the "application of verifiable best practices for the management of forest resources, including woodland and trees, in ways that are ecologically sound, economically viable, socially responsible, and environmentally acceptable and that do not reduce the potential of these resources to deliver multiple benefits").[7] The problem is applying these principles when it is difficult to measure and attribute costs and benefits. Different stakeholders value forests differently. Increasingly, for example, people recognize that real gains could be achieved by the application of reduced-impact logging techniques, and there is growing consensus in the international forest community about parameters to be considered in defining sustainable forest management (partly because of such initiatives as the Intergovernmental Panel on Forests and the Intergovernmental Forum on Forests). Many international and national organizations are in the process of developing criteria and indicators for such management. There is considerable disagreement, however, about the nature and extent of differences in the private and social costs and benefits between conventional and low-impact logging. Broad agreement has yet to be achieved on either the financial viability and the long-term fiscal implications of promoting reduced-impact logging or on the incentives needed to promote reduced-impact logging. These must be worked out country by country.

Given so much close scrutiny and the high perceived risk to the Bank's reputation, the Bank has been reluctant to get involved, not only in the management of timber production in primary tropical moist forests, which the 1993 Operational Policy restricted, but even in the management of secondary tropical forests, which the policy did not restrict. But the Bank could significantly help control deforestation and degradation by helping borrowing countries improve their forest management practices—through improved policies, technologies, and enforcement of regulations—instead of waiting for a consensus on sustainable forest management to develop, especially in the secondary natural forests. The pertinent question is *how* and *how much* will a forest be exploited, rather than *if* it will be, and who will gain or lose in the process.

Country Conditions and Forest Strategy

In the past, forests were viewed as resources to be exploited to facilitate growth, but even some of the earliest analyses of forest use—based on the historical analysis of market forces and ignoring externalities—predicted that forest exploitation would go through several phases, leading to managed forests and tree plantations, as economic conditions improved.[8] Countries in various phases of development are likely to have different incentives to manage forests (Hyde 1999), a conclusion implicit in the Kuznets environmental curve (Panayotou 1995). The Kuznets curve predicts that at low income levels, economic growth will put pressure on forests and increase deforestation, but as income grows, deforestation will stop and forest coverage will increase because of improved government institutions and reduced dependence on agricultural and forest production. Higher incomes lead to increased demand for environmental and other ecological services that place value on maintaining forest cover. This has been the experience of several industrial economies. Some observers challenge this observation, arguing that the income increases said to be needed for such transitions are far too high (Stern, Common, and Barbier 1996). The same observers also note that there is considerable variability around the mean and that it is the median income—which is considerably lower than the mean—that matters for deforestation rates in most developing countries. In any case, environmental awareness is already growing in developing countries, and public opinion about the environment is changing—even in countries with low levels of median per capita income. Many countries have begun applying their own environmental protection policies. The governments of the two largest low-income developing countries, China and India, adopted a policy of no logging in natural forests on their own initiative, India in 1988 and China in 1998, and in Brazil the debate about environmental management has become more active.[9]

Analysis has also focused on differences between the private and social costs and benefits of forests (Shogren and Tschirhart 1999). Because of externalities, without public action, including proactive government policies, market forces would not bring about socially desired outcomes and would therefore encourage deforestation and the loss of biodiversity. And how does one value the irreversible loss of biodiversity, such as the extinction of species? International and national policies, including compensation for cross-boundary benefits, are required if deforestation and loss of biodiversity are to be reduced nationally, and thus globally.

Several questions frame the analysis that follows in this report. How well has the Bank implemented the 1991 strategy? Have forest-rich and forest-poor countries adapted their forest practices to the Bank's 1991 Forest Strategy? What have we learned that makes changes in the strategy desirable or necessary? How do various stakeholders view the strategy?

Bank Group Forest-Related Services and Lending

The World Bank Group and the Global Environment Facility (GEF) together committed $37 billion in 745 approved operations in all sectors during the 1999 fiscal year. Shares of the total commitments to all activities vary greatly among Bank institutions. The World Bank share is nearly four-fifths; International Bank for Reconstruction and Development (IBRD) loans constitute 60 percent of the total, followed by International Development Association (IDA) credits (18 percent); International Finance Corporation (IFC) financing approvals (18 percent); and GEF grants and Multilateral Investment Guarantee Agency (MIGA) guarantees (4 percent each). (World Bank 2000a, b)

Nonlending Services

A Country Assistance Strategy (CAS) is a broad development framework produced by the Bank in collaboration with the country government and other stakeholders and tailored to individual country needs. The main tool used by the Bank's management and Board for reviewing and guiding the Bank Group's country programs, it is also an important benchmark for judging the impact of the Bank's work. Since the mid-1990s, the Bank's emphasis on the "greening of the CAS" has resulted in improved treatment of environmental issues, but the forest sector has not received much attention. Country teams often fail to view sectoral issues, including forests, strategically in country strategies, even in countries where forest sectors are important to the macroeconomy. Even when CASs say the right things about the forest sector, their integration with operational activities has fallen short, except in a few countries. Moreover, the CASs are increasingly developed in close collaboration with the countries and are supposed to reflect their priorities. Some governments' national objectives in the forest sector are different from those of the Bank, leaving little room for Bank involvement in their forest sectors. But there are exceptions. Forest sector issues are addressed in the CAS for Nicaragua, for example, and linked to project design that emphasizes extensive stakeholder participation in project preparation. Treatment of the forest sector in the CAS has also been relatively better in Cambodia, Indonesia, and Papua New Guinea, where forest sector issues have been a part of conditionality in recent stabilization and adjustment lending by the International Monetary Fund (IMF) and the Bank, partly because governments had not been keen to get the Bank involved in their forest sectors.

Economic and sector work (ESW) underpins the Bank's lending operations, informs policy dialogue, responds to country requests for specific advisory tasks, and provides analytical research for the development community. Yet forest-related ESW has declined sharply, particularly since 1995 (Annex C), the 1991 forest paper and GP 4.36 notwithstanding. Sector work on the environment and on natural resource management has increased significantly, but has not addressed forest-specific policy issues. And the quality of the very limited forest sector work has been variable. The Latin

America and Caribbean Region (LCR) has produced by far the best ESW (see Brazil and Costa Rica country studies and the LCR portfolio review) (Contreras Hermosilla and others 2000). Good-quality ESW has also been produced in the Europe and Central Asia (ECA) Region (particularly in Russia) and, more recently, in the East Asia and Pacific Region (see the EAP portfolio review) (Lele and others 2000b). But intersectoral links crucial to the forest sector have generally not received much attention. Although Bank research and lending experience document the effects of agriculture, transportation, and infrastructure projects on rates of deforestation (including their net economic, social, and environmental effects on forest-dwelling people), ESW has not adequately addressed these effects. The Bank can draw on strong ESW in only a few countries if it is called upon to quickly advise governments on forest issues.[1] Strengthening the Bank's analytic base on forests requires much more explicit integration of forest sector issues into the Bank's ESW and more resources need to be devoted to analytically sound forest sector work. Currently, resources tend to be tied to specific investment operations; resources independent of investment operations need to be allocated for ESW about forests.

Forest Sector Lending

This report considered Bank forest-related activities on all public forestlands, including protected areas, and on nonforest lands, including trees on private farms and community lands and all forest products and services emanating from them—in other words, both natural and planted forests, large or small, monoculture or mixed, public, private, or community-owned. Viewed this comprehensively, the 1991 strategy has had a chilling effect on forest sector activities of the World Bank Group, for several reasons. Despite the ambitious objectives of the forest strategy, forest sector issues have been inadequately addressed in the CASs and ESW, and the level and composition of forest sector lending—which is often critical for forest generation and new plantings in forest-poor countries—has been stagnant. Direct forest lending, which has traditionally dealt with forest sector activities, has stagnated. And although forest-component projects have nearly tripled since 1991, they typically address issues associated with the conservation and preservation of natural resources, including forests; they have paid little attention to forest management issues or to external threats to the forest sector. And by themselves, these component activities are not enough to leverage changes in government policies and institutions and to achieve the wider objectives of the 1991 Forest Strategy.

Even though lending for forests (both forest projects and forest components of other projects) has increased 78 percent—from $1.97 billion before 1991 to $3.51 billion after 1991 in nominal terms—forest lending as a share of all lending has remained below 2 percent (figure 2.1). Furthermore, the new mix of forest and forest-component projects may have increased the overall risk of the forest portfolio (see "Performance of forest and forest-compo-

FIGURE 2.1. WORLD BANK FOREST LENDING BEFORE AND AFTER 1991

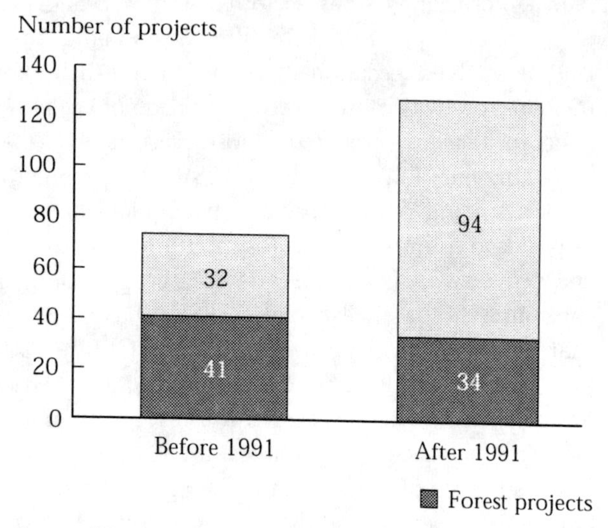

■ Forest projects □ Forest components

Source: World Bank databases.

TABLE 2.1. WORLD BANK FOREST LENDING BEFORE AND AFTER 1991, BY REGION

Region	1984-91 (46 countries)		1992-99 (57 countries)	
	No. of projects	Commitments ($million)	No. of projects	Commitments ($million)
AFR	33	516	23	272
EAP	11	729	35	1,196
ECA	3	45	12	319
LCR	8	254	29	633
MNA	2	69	11	333
SAR	16	361	18	759
All Regions	73	1,973	128	3,512

Note: Includes forest projects and forest-components of projects.
Source: World Bank databases.

nent projects," below). Total forest lending to the 20 countries with threatened tropical moist forests—the focus of the strategy—increased by 36 percent in nominal terms after 1991 (to a total of $1.3 billion), but the projects and components generally avoided working in the tropical moist forests (Annex C).

IFC lending for forest operations declined after 1991, despite an increase in lending to the ECA Region, suggesting that the drop in other Regions that contain tropical moist forests has been even larger. The IFC, in analysis done for this review, was unable to establish if the investments it turned down because of Bank policy in tropical moist forests were financed by others (possibly with less environmental scrutiny than in the IFC). OED and Environmental and Socially Sustainable Development Network (ESSD) consultations for this review suggest that the Bank sent a less than encouraging signal to international organizations and lending institutions about its own involvement in the forest sector; the consultations themselves created new expectations for a more proactive stance by the Bank.

That forest lending in the Africa Region (AFR) also declined (by 47 percent) from its pre-1991 level should be of particular concern, because, next to Asia, Africa has the greatest concentration of poor people, and the forest-dependence of its poor is by far the greatest. At the same time, net lending for forests (figure 2.2) increased greatly for the ECA Region. The 1991 forest paper did not focus on forest issues in the ECA Region, although that Region has by far the world's largest forest area. Nevertheless, uninhibited by the controversy surrounding formulation and implementation of the 1991 Forest Strategy, forest lending increased in ECA by 614 percent as the former Eastern bloc countries started to become Bank members after the fall of the Berlin Wall in 1989 (figure 2.2). The Bank's EAP Region had the largest share of lending after 1991.[2]

Consistency of Project Design with the 1991 Forest Strategy

Project design at entry for both forest projects and projects with forest components has largely reflected the intent of the 1991 Forest Strategy, although some aspects of the strategy have received more attention than others (see figure 2.3). At least 70 percent of all projects were designed to address one or more of the following objectives: poverty alleviation, institution building, participation, and adoption of new technologies. Roughly half of the projects included significant activities in policy reform, forest protection, and forest expansion and intensification. Increasing community participation in forest resource management was especially emphasized in the South Asia Region (SAR) and LCR. International cooperation was addressed in only 32 percent of the projects. Cofinancing activities included donor consultation groups, donor coordination activities, and the incorporation of lessons learned by donors (see the SAR, AFR, and EAP portfolio reviews). The country case studies provide insights into interactions between the Bank and other donors in the forest sector at the country level (see the next chapter). Links within the natural resources management sector were more frequent than those outside the sector in

FIGURE 2.2. NET CHANGE IN BANK FOREST COMMITMENTS

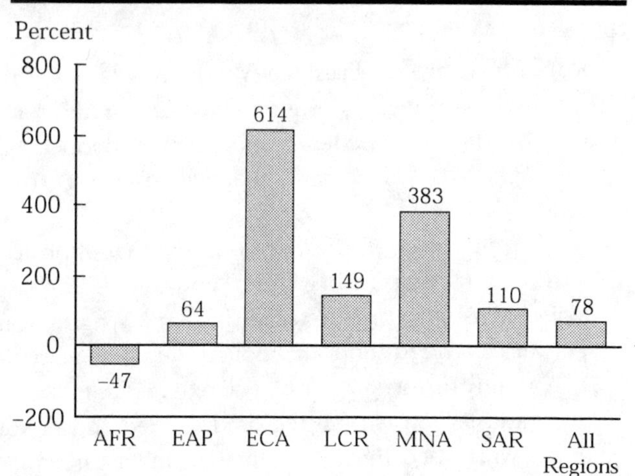

Note: Includes forest projects and forest components of projects.
Source: World Bank databases.

FIGURE 2.3. INCLUSION OF KEY ELEMENTS OF BANK STRATEGY IN FOREST PROJECTS AND PROJECTS WITH FOREST COMPONENTS

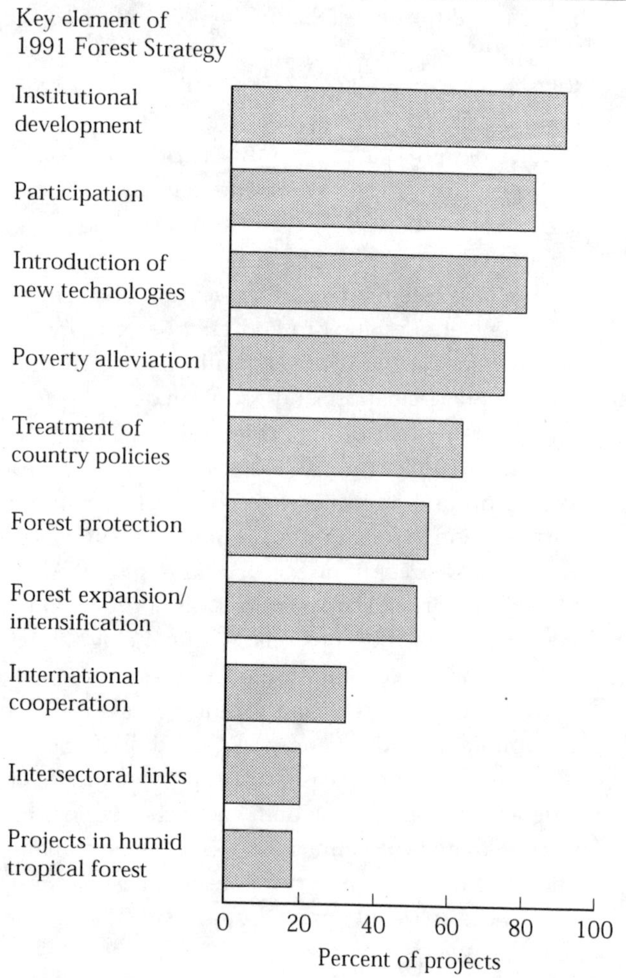

Note: Intersectoral links refer to projects with links outside the natural resource management sector.
Source: Annex C.

project investments. President Wolfensohn's recent initiatives have given additional impetus to efforts to improve the Bank's role in international cooperation (see "Global trends and changes affecting forest policy," below).

The character of forest projects and that of projects with forest components differ considerably. Forest projects, usually prepared by the Bank's agriculture specialists, tend to support production-related activities such as agroforestry research and extension, improved land cadastres, tree planting on degraded and private lands, watershed and forest management, and planning. Forest components, typically prepared by the Bank's environment sector specialists, tend to emphasize "getting the policy and institutional framework right" for conservation in project areas. Forest components address such issues as zoning; the clarification of tenure rights (especially of indigenous populations); the identification, demarcation, and improvement of protected areas; and the enhancement of environmental impact assessment systems in borrowing countries. Forest dwellers often participate in activities related to forest protection and conservation. Conservation issues are often linked to tenure rights of forest dwellers and indigenous peoples. Neither forest projects nor projects with forest components deal adequately with external threats to forest cover or quality or with such mainstream forest sector management issues as concession and royalty policies, public forest enterprises, and the activities of forest ministries and departments. Not enough attention has been paid to improving the capacity of the forest ministries and departments to manage public forests, to setting transparent rules within which forest organizations and individuals (including public servants) can operate, and to building information systems to assess changes in outcomes.

Also, the synergy between forest production and conservation is critical to the design and implementation of an effective forest strategy. Increased emphasis on tree planting for production and productivity growth (including investment in research and extension on public forestlands, watersheds, community lands, and private farms) in the Bank's agricultural and forest sector investments will increase supplies to meet the burgeoning local, urban, and international demand and will meet many of the environmental objectives forests serve. With the notable exception of China, however, tree planting has received little support in Bank lending (see the AFR portfolio review) (Kumar and others 2000c). The Bank has also not done very much to support the enforcement of current forestry laws or to strengthen forest laws and regulations when needed, so little progress has been made in reducing illegal logging or setting up the processes to do so.

Performance of Forest and Forest-Component Projects

The shift in the project mix may have increased the risks of the forest portfolio. The risk ratings of the Quality Assurance Group (QAG) indicate that active forest projects perform better than projects in the agriculture or environment sector, or the Bank as a whole, in terms of risk (see Annex D for details of risk

FIGURE 2.4. PERCENTAGE OF ACTIVE PROJECTS NOT AT RISK

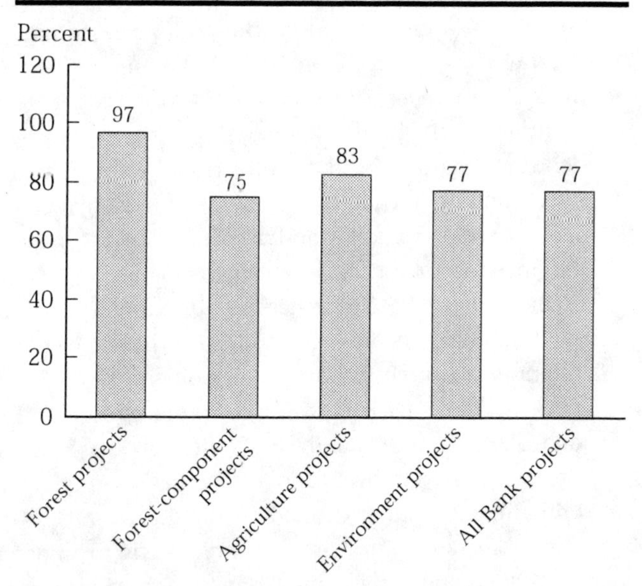

Source: QAG, June 1999.

FIGURE 2.5. COMPLETED PROJECTS: OVERALL SATISFACTORY OED PERFORMANCE RATINGS

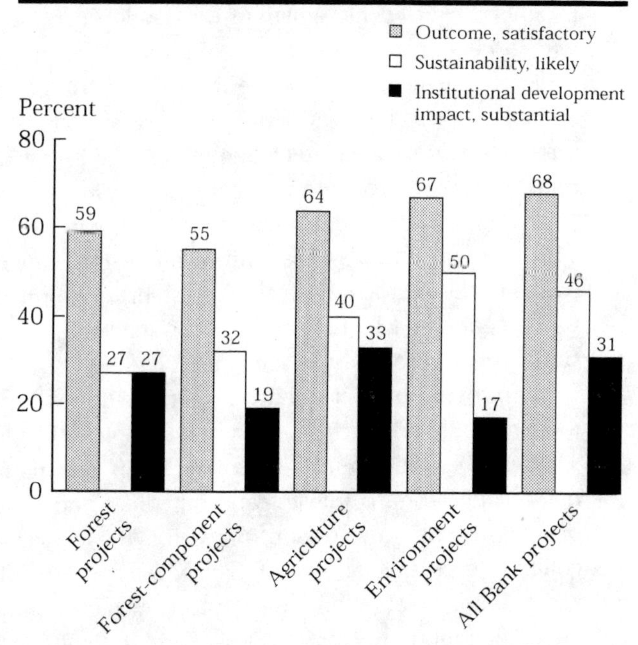

Source: OED databases.

ratings). Forest-component projects, however, are most at risk. QAG evaluated 30 forest and 80 forest-component projects and found that only 3 percent of the forest projects were "at risk," but 25 percent of the forest-component projects were (figure 2.4).

OED's ratings for completed forest and forest-component projects (figure 2.5) are generally much lower than the QAG ratings of active projects. OED evaluated 37 forest and 31 forest-component projects for outcome, sustainability, and institutional development impact (see Annex E). OED's ratings of outcome and sustainability for completed forest and forest-component projects are also generally lower than those for projects in the agriculture or environment sector, or for all Bank projects. OED's ratings for institutional development impact are low in all sectors, but particularly low for environment projects and projects with forest components.

Generally, Project Status Reports inadequately report on progress toward achieving project development objectives. Explanations of forest-related project performance ratings often lack substance. Four kinds of problems are apparent: a weak format, a lack of issue-orientation, inconsistencies, and inadequate follow-up. The format weakness is material, especially because it has not required tracking of compliance with safeguard policies.[3] In some cases, information is vague about the status of actions for achieving development objectives, or even for making progress on implementation. In other cases, ratings of project risk and performance are inconsistent with the explanations given of the project development objective and the project status ratings. There were no managers' comments for 76 percent of the projects. This helps to explain the relatively high disconnect between supervision and completion ratings and between QAG and OED ratings.

There has been some progress in improving the monitoring and evaluation (M&E) of forest-related projects since the introduction of Project Appraisal Documents and logical frameworks. However, substantial challenges remain in crafting well-defined project objectives and verifiable performance indicators that can (a) serve as a basis for agreement and understanding among key stakeholders on what the project is expected to

OED-QAG: The marked disparity in ratings between OED and QAG can be explained in two ways. First, ongoing forest projects may perform better than past forest projects because they have incorporated recommendations and lessons of experience. Second, without effective monitoring of development results, ratings for OED criteria (such as effectiveness, efficiency, and institutional development impact) become clear only at project completion.

achieve; (b) help manage outcomes and impacts within the framework of the intended projects; (c) account for progress toward achievement of the development objective; and (d) address the difference between project achievements at the end of the project and full impact, which may not be clear until some years later. M&E systems do not adequately flag projects with serious implementation problems, partly because systemic problems inhibit the establishment and functioning of the M&E system. OED's observations about the status of M&E in this sector review are consistent with previous OED reports on M&E. When M&E is inadequate, well-staffed supervision missions are vital.

Improving project quality at entry requires laying more groundwork during project preparation and design. This can be done by clearly linking goals and performance indicators to impacts and outcomes and, in the process, firming up responsibility for the collection, collation, reporting, and use of data. M&E and providing access to information at the sector level or higher should be an integral part of project design and of the financial plan. Systematic efforts are also needed to improve the quality of information on the extent of the forest cover and the amount of employment the project has generated.

Policy and Institutional Reform

The 1991 forest paper identified policy reform and institutional strengthening as key principles for future Bank involvement in client countries' forest sectors. In addition to policy dialogue based on ESW, the Bank has used two approaches: project-based reform and adjustment lending-based reform. The effectiveness of project-based policy reform depends on correctly identifying issues and persuasively demonstrating their relevance. The forest sector is complex and multidisciplinary, so the choice of policy issues to address in projects is critical. Moreover, if the assumptions underlying reform issues do not fit the circumstances in a given country, the policy advice may not lead to the expected outcomes and its implementation may be uneven. If a country's domestic interests conflict with Bank advice, even after decades of lending for policy and organizational reform, achievements may be negated, as they were in Kenya.

Attention to policy and organizational reform issues has increased in recent Bank projects in the ECA and EAP Regions. (Rozelle and others 2000; Kumar and others 2000a.) Projects in SAR and AFR have given more attention to organizational development than to policy reform. The experience of large countries such as India and China suggests that the Bank often takes a project-by-project approach toward policy and institutional reform, with mixed results, depending on the countries' own commitment to reform. The Bank needs to undertake project lending in the context of a long-term, overarching strategy toward policy and institutional reform.

Policy and institutional reform through project lending tend to command a lower profile and attract less controversy than adjustment lending. The cycle for project lending is generally longer than for adjustment lending, so there is more time for project preparation and design, consensus building among stakeholders, and implementation. Project lending can also deal not only with "stroke-of-the-pen" policy reform (such as the removal of subsidies) but also with reform that requires time for both organizational and human capacity building. Structural adjustment loans, by contrast, are brief and quick-disbursing and, because they are undertaken in a period of crisis, do not allow the time needed for long-term institution and capacity building, or even for consensus building. They do help bring previously neglected policy and institutional reform issues to the

M&E: In some cases, such as China, it is possible to determine precisely the number of trees and areas planted, the extent of tree growth in planted areas, and the environmental impacts of forest projects on soils, water, and carbon. Even in China, however, crucial information on projects' socioeconomic impacts on such things as markets, prices, and incomes is limited, and information about implementation of environmental impact assessments in non-forest projects has tended to be spotty (see Rozelle and others 2000, table A.17). It is also difficult to quantify from Bank documents the extent to which projects have actually contributed to new tree planting. Crucial information on the nature of land tenure regimes (whether public, private, community, or collective) is also often limited or lacking altogether in appraisal and supervision documents. Therefore, China's government has questioned the Bank's re-estimated rates of returns on completed projects, a critical element in discussing the interest rates at which poor households can borrow for forest sector investments.

Institutions: According to the new institutional economics, institutions are the rules of the game—whether political, social, or economic—that shape human interaction. Organizations differ from institutions in the same way that rules differ from players. Rules (policies) define the way the game is played (North 1990). Sometimes what Bank projects mean by institutional reform or strengthening is actually organizational reform rather than (policy) institutional reform.

attention of people at a higher level and can mobilize internal constituencies that already support specific reforms. But conditionality-based lending can also arouse nationalist fervor and elicit criticism of unnecessary Bank interference in domestic affairs, reinforcing resistance, especially from those in government who oppose reform but are responsible for its implementation.

The Bank's leverage to get countries to undertake forest policy reform may appear greater in small, poor countries than in large, richer countries because the small, poor country depends more on Bank resources and is less able to overtly resist external pressures. In practice, however, the effectiveness and sustainability of reform depends on political will, a consensus for reform, and implementation capacity, as experience in Cameroon shows. Adjustment lending used to deploy forest conditionality has shown mixed results to date (as will be discussed in the next chapter).[4]

Environmental Adjustment and Sustainable Development

The World Bank's adjustment lending policy (Operational Directive 8.60, December 1992) explicitly discusses the links between adjustment policies and the environment, including forests. The Bank's support for adjustment lending, which began to take center stage in the early 1980s, increased 65 percent after 1991, amounting to commitments of $57.1 billion for 378 operations, or 30 percent of total lending (Annex C).[5] Despite the shift in lending composition and use of the multisectoral approach recommended in the 1991 forest paper, only four adjustment loans have directly addressed forest issues.[6]

Do typical Bank macroeconomic and sector adjustment loans (with no specific focus on forest sector reforms) address the loans' impacts on forests and the short- and medium-term tradeoffs between economic stabilization and forest sector objectives? And can the Bank use adjustment lending—when its leverage with borrowers increases—to achieve the kinds of reform in the forest sector it is unable to achieve in normal times, as some have argued it should?[7]

In answer to the first question, the review concludes that the Bank has made little progress in addressing the impacts of adjustment lending on the forest sector. The issue is complex because the short-term impacts may be adverse, whereas the long-term impacts—through macroeconomic stabilization, which is essential for sustainable growth—may be positive. Tracking the impact on the ground is even rarer. The Bank does not require environmental impact assessments for structural adjustment loans, but recommends it as good practice. Several loans refer to the establishment of national environmental action plans, but there is no indication of follow-through or that the plans' impact is measured. The lack of an explicit treatment of forests implies that there are no short-term tradeoffs between adjustment outcomes and forests. Yet policies associated with economic crisis and adjustment—such as devaluation, export incentives, and the removal of price controls—tend to boost production of tradable goods, including agricultural and forest products. In doing so, if there are no mitigatory measures, they encourage forest conversion. Moreover, the constrained fiscal situations associated with IMF/Bank stabilization adjustment programs lead to reduced public spending on environmental protection and reduce the forest and environment ministries' already weak capacity to enforce laws and regulations. The situation becomes even more complex when country-specific realities require repeated

Reform: Government commitment is a key determinant of development effectiveness. In a paper on the design of conditionality in adjustment lending, Collier and others (1997, p. 1406) assert that the attempt to buy policy changes actually exacerbates lack of government ownership of policies. If donors price reforms, they buy them and governments sell them, who then owns the reforms? If not the government, the reforms may lack credibility. And if government is to reform, it must design the reform policy, first determining objectives, then choosing appropriate policy instruments. Present donor arrangements undermine this process by specifying the policies governments must adopt to receive aid.

Econometrically analyzing why adjustment programs succeed or fail, using a database of 220 reform programs, Dollar and Svensson (1998) conclude that it is crucial for development agencies to select promising candidates for adjustment support. When a poor candidate is selected, devoting more administrative resources or imposing more conditions will not increase the likelihood of successful reform. They show that although the World Bank devotes more administrative resources to failed programs than to successful ones, those resources have no impact. Because they found no evidence that any of the variables under the Bank's control affect the probability of an adjustment loan's success, they suggest that the role of adjustment lending is to identify reformers, not to create them. Adding more conditions to loans or devoting more resources to manage them does not increase the probability of reform.

structural adjustment packages. It is imperative in such situations to adequately address the environmental and social consequences of adjustment.

In answer to the second question, the review examined four adjustment loans with forest-specific conditions, in Cambodia, Cameroon, Indonesia, and Papua New Guinea. These operations sought to achieve such reforms as replacing logging concessions with forest management concessions, introducing forest management plans, making the concession award process transparent and accountable by introducing tender or auction systems, and lengthening the duration of concessions as a way of increasing incentives for improved forest management. Other reforms included introducing performance bonds and independent inspections to monitor regulatory compliance, replacing log export bans or export taxes with royalties linked to world market prices, and making royalties area-based rather than volume-based, for administrative convenience.

The outcome of these programs is likely to be affected by the coherence of proposed reforms, the extent to which they address prevailing country realities, the quality of the dialogue linked to government commitment, the Bank's willingness and capacity to use a truly participatory approach, and institutional capacity within the countries to implement reform. A history of poor dialogue between the Bank and its client countries about the forest sector explains the Bank's eagerness to use the opening provided by an economic crisis to push reform in the forest sector. But the risks involved in doing so are significant, because the sustainability of promises made under pressure to secure quick-disbursing assistance is far from certain. An approach to conditionality that cannot elicit real commitment (let alone build necessary capacity) lacks credibility. If the Bank is to remain in the forest sector over the long haul, it needs to do three things. First, the Bank and the borrowing countries together need to develop an avowed and broadly known joint commitment to Bank involvement. Second, the Bank needs to develop lending standards based on strong, current policy analysis conducted with the active involvement of nationals. Third, the Bank should encourage the development of institutional capacity and engage in a dialogue with countries to stimulate the countries' will to implement reform. The current Bank approach to forest sector adjustment lacks both a well-established, broadly understood long-term strategy and a commitment of the high-quality staff resources needed to do essential groundwork or to establish rapport with the constituencies (inside and outside the government) that must support reform to ensure its broad domestic ownership, transparency, and accountability. The Bank needs to work toward these goals.

Poverty and Participation

In implementing forest strategy, the Bank needs to recognize explicitly the sector's role in poverty alleviation. Forests play an important role in meeting the direct consumption needs of forest-dependent people. Marginalized populations, ethnic minorities, and women make a living from forest activities through the collection and sale of timber and non-timber forest products; through paid work in planting, harvesting, processing, and marketing forest products; and through providing services related to ecotourism and recreational activities associated with forests and biodiversity. Forest dependence is particularly high in forest-poor countries, where low-end poverty is often concentrated. And the poor can be important guardians of biodiversity, which makes the decline in lending to Africa, where millions depend on forests for their livelihood, a matter of concern. How much a community depends on forest products depends on the condition of the forest, its proximity to the community, access rights, local and external demand, market opportunities, and alternative income-earning opportunities. The poorest households, with little or no agricultural land or livestock, tend to be the most forest-dependent, but they typically have little say in decisionmaking. In addition to improving the prospects

Structural Adjustment Loans: In a World Bank Operational Memorandum dated June 5, 2000, the vice-president of Bank operations issued a clarification of Bank policy on adjustment lending, stating that Sectoral Adjustment Loans (SECALs) "are subject to the requirements of this policy [Environmental Assessment OP 4.01], that is, they are screened and classified as Category A, B, or C operations." The memorandum goes on to note that "Adjustment loans other than SECALs are not subject to the requirements of OP 4.01," but "it is good practice for Bank staff . . . to review environmental policies and practices in the country, take account of any relevant findings and recommendations of such reviews in the design of structural adjustment programs, and identify the linkages between the various reforms proposed and the environment."

for earning a living, participatory development of the forest sector can improve collective decisionmaking, empowering local communities and increasing their social and economic capital. Investment operations like those the Bank has financed using participatory approaches can be helpful not only in building social capital but also in identifying areas of policy and institutional reform that better meet the needs of the local community, increase their voice in decisionmaking, and make poverty alleviation programs more efficient. The consultations for this review, however, revealed that some NGOs oppose *any* Bank involvement in project investments on the grounds that it increases country indebtedness or involves forest ministries or forest departments that are not committed to reforms. Of course, poverty alleviation requires more than participatory approaches. Access to credit, to land for tree planting, and to the regenerated production of timber and non-timber products can greatly increase poor people's incomes. But projects strong on participatory approaches have demonstrated a need for improvement in their treatment of such issues as markets and prices (which can have impacts on the lives of the poor). The OED review was unable to ascertain how forest lending would affect poverty. Baseline data on poverty in forest areas and analyses of the extent to which Bank projects improve employment and incomes—including the relative abilities of forests and alternative land uses to reduce poverty—are rare both in ESW in the Poverty Reduction and Economic Management (PREM) Network and in Bank project lending. But given the high incidence of poverty in some of the Bank's Regions (particularly SAR, EAP, and AFR) and poor people's dependence on forest resources, there is clearly considerable untapped potential for forest sector lending to address the Bank's mission of poverty alleviation.

Participatory, community-based approaches emphasizing empowerment have been evolving since the 1970s. The Bank's 1991 Forest Strategy emphasized greater involvement of local people in the long-term management of natural forests, and forest projects have increasingly adopted participatory approaches in their design. In a majority of projects, both the level and breadth of participation has increased since 1991 (especially after 1994), although these gains vary across Regions, often reflecting the extent to which borrowing countries pursue participatory approaches in their own programs (see Annex C). SAR, EAP, and

Women returning with fuelwood. They walk for an hour to an area where they can find wood. They have to make 2–3 journeys a week. Yatenga province, Kalsaka village, Burkina Faso. Photo courtesy of Still Pictures.

LCR employ more participatory approaches than other Regions. The Bank has played an important catalytic role in operationalizing and scaling up domestic participatory approaches in several countries—especially China, India, and Mexico—that were committed, but lacked financial resources for the training and capacity building needed to put the approaches into practice. Hands-on approaches have been important learning experiences, helping the Bank staff identify policy and institutional reforms crucial for improving incomes and giving more voice to the poorest households. The Bank needs to help governments follow through on essential reforms more than it has.

Plans for making Bank projects more participatory have become more ambitious during appraisal and design, but implementation has lagged, because of six main weaknesses: the inadequate reflection of social, technical, institutional, and political realities in project design; the weak capacity of grassroots institutions; the Bank's failure to consult with key stakeholders from civil society during project preparation and implementation; too little time and too few resources for project planners to develop genuine participatory approaches; insufficient expertise in participatory techniques among Bank staff and consultants; and the selection of M&E indicators that failed to link the project's participatory goals to investments to assess impacts on the ground. The Bank has also been weak in exchanging experiences within and across Regions, except through the ad hoc efforts of individual task managers. The Bank should draw on the resources of the World Bank Institute by linking the development training of borrowers with Bank operations. Because country managers control budgets, the Bank's network has lacked the mandate, instruments, or resources to design and implement a proactive strategy for the forest sector.

Gender

Some of the poorest forest-dependent people are women, but gender considerations have received little attention in the implementation of the Bank's 1991 Forest Strategy.[8] The key to involving women in forest projects is to identify their potential role up front. Equally important, project design should minimize the losses women are likely to have to bear as a result of planned interventions—such as having to walk longer distances to gather fuelwood so that nearby forests can be protected. Involving women in microplanning is an ideal way to deal with these issues. Such approaches call for drawing on the resources of organizations that specialize in women's participation and are capable of instilling confidence in rural women, understanding and addressing their concerns, and ensuring that they play an active role in decisionmaking. But GP 4.36 notwithstanding, the Bank-wide share of projects that state that gender concerns are their primary objective is less than 0.5 percent. Project design and implementation has a long way to go in genuinely involving women. OED's review of implementation in India shows that although women are encouraged to participate in committees, there is little monitoring of their effective presence. Meetings are scheduled to suit men, at times when women tend to be cooking. Even when women attend meetings, they rarely participate. The shortage of women qualified to implement projects is also a problem. Given the complex cultural challenges of getting women involved in poverty-oriented forest projects, the Bank should make extra efforts to engage nationals of borrowing countries experienced in handling such challenges.

Safeguards

One important commitment of the Bank's 1991 Forest Strategy was to reduce the negative impact of Bank activities on the world's forests and the people who depend on them. This was to be achieved by using a multisectoral approach that included environmental impact assessments and efficiently applied safeguards to minimize projects' environmental and social damage.[9] Safeguard policies include the 1993 Operational Policy on forests. These policies are now better incorporated in project design, but are not systematically monitored for quality at entry and supervision. According to QAG, the Bank's performance in monitoring compliance does not meet the 100 percent quality standard expected for safeguard policies (World Bank 1999, p. 15). Current data systems do not help staff identify and anticipate potential indirect and long-term forest problems that arise from projects in sectors such as transportation and infrastructure. These problems can lead to substantial changes in land use with negative impacts on forests—although there are several positive examples of how environmental impact assessment is monitored—for example, in Indonesia's infrastructure sector (see Gautam and others 2000). Bank documents on large public infrastructure projects generally contain little information on whether forest and biodiversity issues are relevant, especially when these concerns are not related directly to project outcomes and sustainability. Even where these issues are addressed, their treatment is often superficial—for example, reporting trees cut to build a road rather than assessing the likely indirect effects on forests caused by increased road access. But the Bank's public accountability for safeguards is improving.[10]

There has been considerable progress since OP 4.36 was issued in 1993. Safeguards for forest and forest-component projects *at entry* do address topics related to the protection of forests and forest dwellers. Bank projects have responded to the interests of stakeholders by incorporating planning and implementation tools—including participatory planning, village plans, ecological zoning, demarcation and land titling, and indigenous reserves—in project design. Moreover, Bank projects in most Regions have been cautious about carrying out production activities in tropical moist forests, focusing on improved forest management plans and on the testing and introduction of suitable technologies. Projects with significant tree planting components have generally not been located in or near tropical moist forests.

Implementation of safeguards in forest and forest-component projects is much more difficult. Most difficulties arise in projects that address issues of property rights and the distribution of forest resources (such as ecological zoning, or the creation and demarcation of indigenous reserves or protected areas). As a result of safeguards, vulnerable stakeholders have been incorporated in project design, but other key stakeholders, including those most likely to cause harm, are often not consulted, usually to avoid criticism that the Bank is being too "cozy" with them. The ways safeguards have been applied can thus contribute to conflicts among interests competing for forest resources—including conflicts between the indigenous poor and the non-indigenous poor, between the powerful and the poor, and others—without helping to establish transparent rules to hold all accountable and to monitor performance. Also, the Bank's universal safeguards often conflict with national legislation and regulations. The Bank does not have the capacities it needs in the right places to deal with these complexities, so task managers often try to avoid projects that are likely to be subject to NGO scrutiny and charges by the borrowing government of interference in sovereign issues (see Annex G). Early attention to safeguard policies and to the development of country-wide capacity (to ensure domestic ownership of a project) is critical to project success. But the "do no harm" principle behind the safeguards has unintentionally increased transaction costs and the reputational risk of investing in projects that could have adverse impacts on forests and the people living in or around them.

The increased number of environmental impact assessments (EIAs) since 1991 reflects heightened attention in Bank projects to the indirect environmental impact of projects (Annex C). With some notable exceptions, however, these assessments have generally not focused on issues critical to a particular project, have typically come too late in the project design process, and have tended to be superficial about forest-related biodiversity issues and short of analysis of indirect and regional impacts. Even where assessments have been satisfactory, it is unclear from available documentation whether recommendations have been incorporated into project design and implementation. The Bank has recently recognized these weaknesses, however, and has improved the design of the supervision form so that implementation can be tracked and monitored.

Links with the Private Sector

Bank investments in the forest sector have done a poor job of fostering private sector participation. Public-private partnerships in the forest sector of developing countries are generally limited. Wood industries in these countries are underfinanced and inefficient, and the lobbying efforts of powerful interest groups have influenced public policies about royalties, concessions, and trade that benefit some parts of the private sector, at the cost of economic efficiency, social equity, and environmental sustainability. Public and private sector research and extension in forest sector activities have often lagged. The combination of weak support services and heavier regulation of the forest sector has created incentives for land conversion from forest to alternative uses. However, new opportunities are emerging for increased private investment in the forest sector in commercial plantations, seedling production, reduced-impact logging, ecotourism, certification, and supervised timber harvesting in forest reserves. Governments and the private sector in borrowing countries appear to be eager for greater Bank Group participation in the modernization of their forest industries and in increased support for silvicultural research and extension.[11]

Tree nursery, Annapurna Conservation Area Project (ACAP), Ghandrung, Nepal. Photo courtesy of Still Pictures.

The Bank is currently involved with the private sector in the regulation of forest concessions (Cambodia 1999), the promotion of certification and forest management (Honduras 1999), and the privatization of government lumbering operations (Croatia 1997). There is ample scope for improving private sector participation by sharing information and exchanging lessons across

Environmental assessment: The World Bank assigns an environmental assessment (EA) category to all its lending operations based on the nature of the project. These categories include full EA (A), partial EA (B), no EA required (C), free-standing environmental project (D), and to be decided (T). After 1991 the number of projects that required full EA increased 871 percent and the projects that required partial EAs, 306 percent. Interestingly, the number of projects with EA undecided declined 73 percent. Free-standing environmental projects also declined, about 20 percent.

projects, countries, and Regions. Whether private sector activity increases will depend mainly on an improved investment climate in the forest sector, the appropriate valuation of forest resources, the establishment of clear and easy-to-implement rules and regulations, and effective government monitoring and supervision. The IFC review found that the private sector has no incentive to invest in forest development as long as a cheap supply of timber is available from publicly owned forests through legal or illegal logging. The Bank Group could work with governments and the private and voluntary sector to support community-based forest operations and to develop progressive, practical policies and information-dissemination mechanisms that ensure the efficient and sustainable use of natural resources.[12]

Biodiversity and Protected Areas

The 1991 forest paper's call for attention to biodiversity, not just to timber, has been answered partly through an increase in Bank financing for biodiversity. Both the Bank and the Global Environment Facility projects are helping to expand protected areas for biodiversity conservation in several countries. Joint Bank-GEF projects have also tackled important regional initiatives, such as the Mesoamerican Biological Corridor. However, the overall scale of interventions has been insufficient to significantly affect global biodiversity.

There may also have been too much focus on community conservation initiatives and not enough on such major threats as pressures on buffer zones, poor logging practices by concessionaires, road construction, and weak in-country ownership.[13] Governments have not adequately enforced the existing laws, rules, and regulations designed to protect against biodiversity loss, and projects have not sufficiently helped improve enforcement. The policy against Bank financing for commercial logging in tropical moist forests may have hindered even GEF support for promising experiments in forest management by local communities and the private sector. Weaknesses in the monitoring of environmental and forest safeguard applications in infrastructure projects, with often large but indirect impacts on biodiversity, also need to be addressed. The development community should use its influence to persuade governments to live up to their international commitments on the environment, but it is unclear if biodiversity of global importance can be saved without (a) establishing clear global priorities for its protection; (b) adequately analyzing the causes of real threats to its loss; and (c) providing assured, long-term financial support for its protection, through arrangements such as trust funds; and d) payments for environmental services by the gainers to the losers.

Bank Group Activities

International Finance Corporation (IFC) operations are typically post-harvest processing activities that use timber or its by-products and are expected to have a direct impact on forests. The IFC approved investments of $578 million in 65 forest operations in 29 countries during the 1992–98 period, a 17 percent decline in commitments from the period 1985–91. After 1991, IFC forest sector lending increased in all Regions but LCR and EAP, which contain much of the world's tropical moist forests.

In evaluating the implementation of the 1991 Forest Strategy in IFC's projects (Annex J), the Operations Evaluation Group (OEG) found that the intention of engaging the private sector in sustainable forest management was unrealistic because more than half of the IFC's projects use government-owned forests, and

Biodiversity: From 1991 to 1999 the Bank undertook implementation of 162 GEF projects; 44 were forest-related, with total commitments of $370 million in GEF funding. All dealt with biodiversity. They involved Bank cofinancing of $181 million. Contributions of donors and national governments brought total project costs to $1 billion. It is difficult to determine how much of this was incremental aid to developing countries rather than the diversion of aid resources from other activities. See World Bank 2000b.

Protected areas: Developing countries have substantially increased the amount of their protected areas but often have difficulty reaching IUCN standards of full protection because the livelihoods of many poor and indigenous communities are based in these areas. Converting protected areas to a higher status (for example, from sanctuaries to complete protection) creates intense conflicts. Many Bank and GEF efforts to expand protected areas have not been complemented by assistance to governments to help raise fiscal resources or undertake cost recovery to better finance the protected areas. (There are exceptions, such as the Bank advice through ESW in Costa Rica and the Minas Gerais project in Brazil.) There is much less agreement in the international community about the merits of helping developing countries make their protection efforts more effective than about adding to newly protected areas to meet globally established targets. Since land values may increase, establishing protected areas may be more costly later, so some argue that as much land as possible should be brought under protection now, regardless of effectiveness.

private operators have few incentives to work with local people, special-interest groups, or forest dwellers in forest areas the concessionaires neither own nor contractually control. Moreover, because the stumpage fees paid are in some cases lower than the real cost of managing the forests sustainably, the private concessionaires have a financial disincentive to encourage the more costly option of sustainable forest management. The IFC encourages its projects to obtain forest certification, but does not make it mandatory. (Forest certification comprises both certification that a forest is sustainably managed and eco-labeling that enables consumers to recognize products made of timber from sustainably managed forests.) From the perspective of the IFC's clients, certification of forest management entails cost on a regular basis and is currently expensive, creating a dilemma for policymakers and forest product companies. At the same time, eco-labeling is a revenue-generating activity that permits premium pricing for differentiated products. Eco-labeling is viewed as a marketing tool for exports mostly in developed countries, and hence currently often irrelevant to forest-based projects whose outputs are sold in domestic markets. Because eco-labeling relies on brand awareness, it is also irrelevant for projects whose outputs are traded as commodities. The recent proliferation of forest certification systems has created an unhealthy rivalry for "membership recruitment." (See Chapter 4 for further discussion of these issues.)

According to OEG's review, the 1991 forest paper led to fundamental changes in the IFC's project selection, processing, and monitoring. The strategy's emphasis on the ban on commercial logging in tropical moist forests had a chilling effect on IFC activities. To avoid any association with deforestation, the IFC made a conscious decision to screen out such operations completely, and has subsequently turned down several proposals as a result. The IFC has not been able to establish if the projects it turned down were financed by others but, in its mainstream operations, the IFC has not approved a single forest-based investment in tropical moist forests since the 1991 strategy became operational, although two small investments—approved under streamlined review procedures through the IFC's small- and medium-size enterprise (SME) facilities—financed a company engaged in transporting logs harvested under concessions in tropical moist forests. (These two investments were consistent with the letter of the forest strategy but digressed from its "spirit and intent," and the environmental review procedures in place at the time failed to detect the inconsistency.) Two other small investments were in companies whose nonproject operations, or those of their sponsors, involved logging in tropical moist forests. IFC has helped establish project-owned, large plantations, particularly in Latin America and Asia, and ensured that those companies' forest operations rely on government forests and are carried out in a sustainable manner, in accordance with good industry practice. The IFC has also supported several projects that rely on wood wastes, wastepaper, and other recycled materials, thus contributing to the conservation of forest resources. The 1991 strategy provided no guidance about projects using temperate and boreal forests, which comprised all of the IFC's forest-based operations approved since 1991.

The Multilateral Investment Guarantee Agency (MIGA) was established in 1988 to provide political risk insurance (guarantees) for new foreign direct investment. It does not participate in project financing or lending. MIGA's Convention does not envisage it having a role in policy interventions in host countries. Until 1999, MIGA had no direct involvement in the forest sectors of its clients. By December 1999 its portfolio included only two guarantees in the forest sector, one for an existing pulp and paper mill in the Svetogorsk region of Russia, and another for the rehabilitation of an existing cocoa plantation in Côte d'Ivoire. MIGA also has contracts of guarantee for four projects located in forested areas: three mining projects (in Colombia, Guyana, and Papua New Guinea) and one ecotourism project (a rainforest tram in Costa Rica). All four projects required full environmental impact assessments before approval. MIGA's proposed guarantees are evaluated for compliance with its own environmental guidelines and the forestry and natural habitats safeguards of the Bank Group. On particularly sensitive projects MIGA uses various means of compli-

GEF Logging Policy: GEF's Interim Guiding Principles for Projects Associated with Logging go far beyond the 1991 Forest Strategy. They indicate that GEF will not support logging in *any* primary forests, not just tropical moist forests, and that GEF financing will not be used to meet baselines for pursuing sustainable forest management, the cost of forest certification, of improving timber harvesting methods to meet Forest Stewardship Council/International Tropical Timber Agreement criteria, or to finance reduced-impact logging, among other things.

ance monitoring, including independent environmental audits and its own monitoring activities. Two of the three mining projects located in forested areas were subject to MIGA's own environmental monitoring during the past fiscal year.

GEF Activities

The Bank is one of the implementing agencies for the *Global Environment Facility (GEF).* This gives the Bank a unique opportunity to "mainstream" biodiversity conservation in borrowing countries' environmental strategies. Since its inception in 1991, the GEF has financed 44 forest projects through the Bank. GEF grants for forest projects amount to $370 million, or slightly more than 10 percent of the Bank's commitments to the forest sector, and reflect about 26 percent of the GEF portfolio. Eleven of the GEF forest projects are jointly financed, involving Bank commitments of $181 million. (For a detailed discussion of GEF activities see World Bank 2000b,)

A review of GEF activities concludes that the Bank's partnership with GEF has allowed it to pursue aspects of the 1991 Forest Strategy that might not otherwise have been possible, since few countries are able to borrow IBRD or even IDA funds for biodiversity conservation (see the next chapter). But it would be difficult to show that these relatively modest interventions in the forest sector have so far significantly helped to mainstream biodiversity issues into Bank country or sector dialogues or to address the biodiversity challenge. Mainstreaming long-term global concerns about biodiversity into the Bank's forest sector lending might be ambitious, considering the many pressing economic and social concerns the Bank's borrowing countries face.

There is a greater scope for grants that focus on demand-driven biodiversity conservation and improvements of the countries' own strategies to address threats to biodiversity of major importance. But this creates the dilemma inherent in the GEF's mandate of supporting only activities of global importance.

The review recommends expanding the partnership of the Bank, the GEF, and the private sector, with more innovative approaches and greater attention to economic, livelihood, and sustainability issues.

The Country as Unit of Account in the 1991 Forest Strategy

Six country studies (of Brazil, Cameroon, China, Costa Rica, India, and Indonesia) reviewed implementation of the 1991 Forest Strategy in Bank operations to find out how non-Bank stakeholders in those countries viewed the Bank's strategy and its implementation.[1] Although the case studies cover only 6 of the Bank's 181 member countries, those 6 countries account for 44 percent of the world's population, 27 percent of the world's forest cover, 31 percent of the Bank's lending, 59 percent of the Bank's forest lending, and 45 percent of the Bank's forest-component lending. Three of these countries (Brazil, Cameroon, and Indonesia) are forest-rich, and three (China, Costa Rica, and India) are forest-poor.

The reviews found that:

- The forest-poor countries seek World Bank support for tree planting or regeneration to a greater extent than do forest-rich countries, which frequently shun Bank involvement and show less commitment to policy and institutional reform.
- The objective of improving tree cover and meeting the basic needs of the poor has also been better implemented in the forest-poor countries than in the forest-rich countries.
- The interests of the local and global community largely coincide in forest-poor countries, but their willingness to borrow on IBRD or even IDA lending terms is in question.
- Forest cover is stabilizing or increasing in the forest-poor countries, but degradation of publicly owned natural forests is still a serious issue.
- Both forest-rich and forest-poor countries have increased the areas declared as protected, but management of these areas is a serious problem in all countries because of insufficient public funds.
- Independent of the World Bank strategy, scarcities have brought about conservation-oriented policies in the forest-poor countries and the forest-poor regions of forest-rich countries (for example, southern Brazil).
- To forest-rich countries, their "abundant" natural forests represent an important source of income generation, employment, government revenues, raw material, and land for alternative uses. Forests serve important development purposes, but exploitation of forest resources is often to be carried out in a manner that is environmentally unsustainable and socially inequitable.
- In forest-rich countries, the interests of the local and global communities tend to strongly diverge. Without reform to cover the opportunity

cost of conservation, this divergence puts the Bank in the difficult position of promoting a dialogue on forestry strategy that can reconcile local and global interests.

- However essential policy and institutional reform is in all countries, the impact of reform tends to be highly location-specific, regardless of the country's forest endowments.
- The Bank's leverage on policy reform has been more limited than the 1991 forest paper assumed it would be. That leverage has been often more effectively exercised with long-term involvement in solving practical problems through policy analysis and investment operations, which helps build confidence and effective partnerships.
- Increased environmental consciousness, even in the forest-rich countries, provides more opportunities for Bank involvement in forest production, development, and conservation activities.
- The development of domestic financing mechanisms and constituencies for forest management offers promise for national conservation efforts, but that promise remains unrealized. Bank financing can support conservation consistent with national priorities and objectives. International mechanisms and global financial transfers involving grants—not the deployment of the limited Bank loans and credits the 1991 Forest Strategy seemed to anticipate—are needed to conserve forests of global value.
- Investment operations based on grants or highly concessional terms can substantially improve the synergy between environmental protection and development through forest production.

Forest-Rich, Forest-Poor, and Transitional Countries

Brazil's tropical moist forest—covering 3.7 million square kilometers and representing one-third of the world's remaining rainforest stock—is the world's largest and richest (figure 3.1) (Lele and others 2000c). Brazil also has 5.5 million hectares of the developing world's most productive and economically competitive forest plantations. Population densities in the rainforest are very low, but great disparities in income and wealth in Brazil encourage migration to the Amazon, creating pressure to use forestland for agricultural expansion and other activities. Forest loss in the Amazon averaged 13,000 square kilometers annually in the 1990s, substantially varying year by year.[2] The reasons for the year-to-year variation in these national estimates are ambiguous, and explanations range from forest fires to macroeconomic difficulties and adjustment. Brazil's *Cerrados* region and biodiversity-rich Atlantic forest are much more endangered than the Amazon, but attention to the Amazon eclipses attention to conservation in other areas. Subsidies for agricultural expansion (to which deforestation in the Amazon was attributed in the 1980s) have declined, but other factors still contribute to deforestation, including agricultural expansion, strong urban and industrial demand for wood products domestically, and investments in extensive transportation networks. Globalization, trade liberalization, currency devaluation, and technological advances have made agriculture and agricultural exports more profitable. At the same time, devolution of power to the state and local levels, the growing economic and political influence of logging and agricultural interests, and forest revenues' increasing value to municipalities and state governments have

Forest-rich and forest-poor: Forest-rich and forest-poor countries were defined by the percentage of total land area reported under forest rather than the absolute size of their forests. Large areas in forest-rich countries can be forest-poor, and forest-poor countries can also have forest-rich regions. The breakdown into forest-rich and forest-poor categories can shed light on the impact of differences in resource scarcities on forest valuation, incentives, and perceptions within and among countries about forest conservation. Abundance or scarcities together trigger policy and institutional behavioral responses that lead to changes from forest-rich to forest-poor or the reverse. Forest-poor southern Brazil has many more pro-environmental policies than the forest-rich north. Costa Rica, once forest-rich, is again in transition, from forest-poor to forest-rich. It was categorized as forest-poor because its forest policies have more in common with the policies of forest-poor countries than with those of the forest-rich. China is similarly undergoing a major transition. Although both countries are increasing their forest cover, and planted forests serve many environmental functions, some loss of biodiversity is not recoverable.

Public forestland: Recorded public forestland in Costa Rica and India may be stabilizing, but estimates of land under forest cover are disputed in all countries as there is de facto encroachment on, and degradation of, public land. Much of the increase in forest cover in China, Costa Rica, and India has come from trees planted on lands outside the public forest areas. In China, however, significant planting has been done on degraded state, collective, and community lands, as well as on individual farms.

intensified political and economic pressure to use the country's forest resources for development. The focus of the Bank's Operational Policy for forestry on ensuring a government commitment to sustainable forest management before Bank investment and on the conservation of tropical forests (see box 3.1 and Annex I) has proved ill-suited for dealing with the complex pressures to develop Brazil's forests. Bank involvement in Brazil's forest sector has thus been limited except through the Rain Forest Trust/Pilot Program (PPG-7) that is piloting a number of small operations, despite the lack of a government policy for the forest sector (box 3.1). The debate on the future of the Amazon has grown vigorous as the government has been formulating a forest policy, scheduled to coincide with the country's five-hundredth anniversary in 2000.[3] The participatory process being used has engaged a variety of stakeholders. Public opinion favoring protection of the Amazon is stronger than it was before, but it is too early to predict the outcome of the debate and its consequences for the Amazon. Brazilian authorities and the private sector have also expressed strong interest in modernizing the forest industry in the Amazon and in developing plantation forests in the south as a follow-up to a successful project in the state of Minas Gerais.[4]

Indonesia's rainforest, also rich in biodiversity, is second only to Brazil's in size. About 78 percent of its territory is recorded as publicly owned forests, although the precise extent of the forest is a matter of much debate (Gautam and others 2000). The high population density of Java compared with that of the Outer Islands has led Indonesia to encourage both migration to, and exploitation of, its forest resources. Indonesia has emerged as one of the largest forest industrial complexes in the developing world and the largest exporter of tropical forest products. Indonesian policies, believed to have been influenced by the "KKN system" (corruption, cronyism, and nepotism), have recently come under heavy criticism by the Bank and other analysts, both domestic and international.

The Bank supported the country's now-notorious transmigration projects until the mid-1980s to relieve population pressure on Java. The Bank also contributed to smallholder tree crop production, which has been both hailed as one of Indonesia's success stories and criticized by some because of the resulting forest conversion and indigenous communities' loss of access to land. The two Indonesian forestry projects financed by the Bank in the 1980s were fully consistent with the intent of the 1991

FIGURE 3.1. RELATIVE FOREST AREA IN FOREST-RICH AND FOREST-POOR COUNTRIES

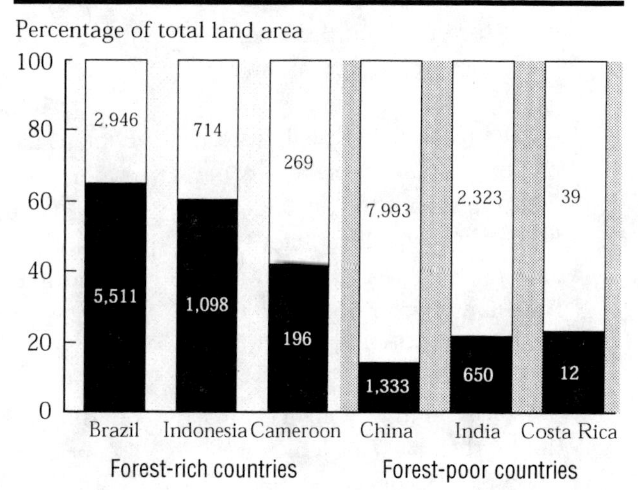

Note: Reported amounts, 1995–96.
Source: Whiteman, Brown, and Bull 1999.

Forest Strategy, focusing as they did on policy and institutional reforms. However, the government asked the Bank to withdraw after a blunt Bank report in 1993 expressed concern about the rapid rate of deforestation in the tropical forests and the role policies and institutions might have played in it. The Bank did not make the staff report public, nor did it pursue reform until the economic crisis in 1997 offered an opportunity to introduce forest sector conditionality in the IMF-Bank stabilization/adjustment packages. By mid-2000 there was a marked improvement in the consultative process among key stakeholders on forest issues in Indonesia, which was noted in OED's case study of Indonesia's forest sector (Gautam and others 2000), but government commitment to reforms remains unclear.

Cameroon's closed-canopy forest covers an estimated 15.5 million to 20 million hectares (Nssah and Gockowski 1999). The country's forests (and biodiversity) represent a major part of the Congo Basin, which accounts for 80 percent of the remaining tropical moist forests in Africa. Cameroon's forest sector experience is strongly linked to developments in the agricultural sector and in the country's political economy. Agriculture's low productivity, combined with increased food demand, has made expansion of the cultivated area a leading cause of deforestation. International logging companies dominate the development and export of forest resources. The poor performance of agricultural projects and Cameroon's

BOX 3.1. BRAZIL: A CASE STUDY IN DOMESTIC PRIORITIES VERSUS INTERNATIONAL OBJECTIVES

Since 1991 the Bank has moved away from direct involvement in forest-related projects in Brazil, instead playing a smaller role as a coordinating agency for the $300 million Rain Forest Trust/Pilot Program (PPG-7) to conserve the Brazilian rainforest (funded by international donors and the GEF).

A recent independent evaluation of the PPG-7 by a blue-ribbon panel concluded that many interesting ideas have been piloted at the micro level (for example, in agroforestry and fire prevention), but that the PPG-7 failed to articulate the program's strategic objectives. The review criticized the program, the World Bank as its coordinator, and other donors involved for failure to agree on a program strategy, for the participants' inability to address fundamental program issues, for complex project design and financing plans, for weak program management, and for slow coalition building with Brazil's civil society and private sector.

Acknowledging that PPG-7 as conceived was almost completely externally driven and had the exceptionally ambitious objective of containing deforestation in the Amazon, the external panel stresses that stronger government ownership of the program is essential for its future success. Positive steps have been taken by getting Brazil's government actively involved in the program's management.

The program has not met its original ambitious objectives, but it has demarcated more than 39 indigenous lands, some extractive reserves now serve as models for conservation and development, and 160 demonstration projects in natural resource management have been carried out. The result has been a stronger relationship among civil society, Brazil's gov-

deteriorating economic condition in the late 1980s prompted the Bank to abandon rural development and virtually all project lending in favor of adjustment lending. In the process, the Bank missed an opportunity to promote rural development. Success has also been limited for the Bank's adjustment lending operations and forest sector reforms for several reasons—among them, the poor sequencing and phasing of reforms, the lack of political will, weak institutional capacity, and insufficient consultation with domestic and international stakeholders, especially the private companies that have resisted reforms suggested by the Bank. The Bank's dialogue with the government and key stakeholders is said to be improving, some legislative and regulatory changes appear to be taking place, and the Bank's resolve to remain involved in the sector is reflected in an adjustment loan and a forest-sector-related agricultural adjustment loan currently under preparation. It is too early to determine either government commitment or what impact these initiatives will have.

What is happening in forest-poor China (Rozelle and others 2000), India (Kumar and others 2000a), and Costa Rica (Velozo and others 2000) contrasts sharply with events in forest-rich Brazil, Cameroon, and Indonesia. China, India, and Costa Rica now have progressive forest policies, and the Bank's efforts have been driven largely by demands articulated by the three countries' national governments. In recent years their overall forest cover has stabilized or improved, although natural forests continue to degrade and disappear. A third of the world's population lives in China and India, including more than half the poor people in the world (those earning less than a dollar a day)—many of whom, especially ethnic minorities, depend on forest resources

Transmigration in Indonesia: A 1994 OED review of the Bank's transmigration projects in Indonesia concluded that the projects had largely succeeded in achieving their narrowly defined resettlement objectives, and most projects had even had beneficial impacts on the welfare of the settlers. The Bank's environmental guidelines were issued after the projects had been appraised. Although the Bank often correctly identified potential negative impacts and proposed mitigation measures during appraisal, follow-up during implementation was weak. Individually, adverse environmental impacts at each site might not have resulted in a major loss of forest or biodiversity; only when the projects are viewed collectively can their serious and unmitigable impacts on the forests be appreciated. The OED review also noted that although the settlers had benefited from the programs and had settled into their new environment, the program had a major negative impact—probably irreversible—on the indigenous peoples, particularly the Kubu, who depend on the forest for their economic and spiritual livelihood (Gautam and others 2000).

ernment, and the Bank. The PPG-7 has also helped to stimulate an active interest in certification and biodiversity issues.

The World Bank/WWF Alliance (box 4.1) has also been driven by two external institutions. The president of Brazil pledged to set aside an additional 25 million hectares of forest for protection, but progress on doing so has been slow. The president's pledge generated widespread internal debate and criticism from several nongovernmental organizations. The financial crisis and budget cuts of 1999 raised additional questions about relative priorities for development and environmental expenditures. Implementation of the alliance pledge has since been reinvigorated with a $35 million GEF grant for preparation efforts on a larger policy- and strategy-oriented program (under discussion between the World Bank and Brazil as this report is being completed). But the government's fiscal austerity program, agreed to with the IMF, considerably limits its desire to undertake new investments.

Several questions remain: will the international community help Brazil preserve the rainforest for the global benefits of preventing climate change and conserving biodiversity, as the 1991 forest paper articulated? And will Brazil demonstrate the political will to make long-term investments in conservation on the grounds of international externalities, when the private sector and some municipal and state governments have strong incentives to deforest and the national government has a strong incentive to bring the macro-economy back on track? If so, how? (See Lele and others 2000c for details.)

for their livelihood. And yet, combined, the two countries have only about 6 percent of the world's forestland (0.1 hectares in China and a meager 0.007 hectares in India per capita). China has 30 million hectares of plantations that help compensate for the loss of natural forests. The Bank is helping to establish another 3.9 million hectares of forest plantations on collective, community, and individual farmers' plots through its forest and forest-component projects. It has also supported the establishment of horticultural trees on a significant scale. Carbon sequestration from the Bank-financed tree and forest plantations is estimated to represent roughly 25 percent of China's annual carbon emissions. India has one of the world's largest livestock populations, which depends heavily on forestlands for grazing. India also has the largest program of participatory forest management for forest regeneration the Bank has ever supported. In the 1990s the Bank did not follow through on the social forestry program of the 1980s, which promoted substantial tree planting on community and private lands, but agricultural intensification has helped relieve pressure on forestlands in both China and India. The Bank has contributed substantially to agricultural intensification through investments in irrigation, fertilizer, research, and extension, reducing the rate of forest conversion.[5]

Costa Rica, a small, middle-income country—with 3.5 million people and a per capita GNP of $2,680—has a total forest cover of 1.2 million hectares. It also has one of the most progressive forest policies among developing countries. Legislation has supported such important initiatives as the "polluters-pay principle," financed by a 5 percent tax on fuels, part of which is used as payments for environmental services to those who plant trees or support other forest activities. Costa Rican laws require certification of good forest management. Carbon Tradable Offset Certificates, which were developed for trading carbon internationally, serve as a model for trading other environmental services. Costa Rica has also been active in protecting biodiversity and in bioprospecting (identifying commercially or medically useful chemicals in living organisms). Fortunately Costa Rica has well-established property rights, and government policy has favored incentives for private owners to increase forest cover. Costa Rican forest owners have strong organizations that give them technical support for reforestation, forest management, and conservation. Costa Rica's progressive forest policies would not have been possible without a strong network of governmental, nongovernmental, and private sector organizations capable of adapting to policy innovations.

The Bank has achieved its greatest impact in Costa Rica through its ESW, which influenced Costa Rican forest policy, even though the Bank did not finance any investment operations in the sector, and even though Costa Rica was unable to mobilize many international transfers for environmental services. What happened in Costa Rica's forest sector shows that if a country is committed to conservation activities, the Bank's ESW can have considerable impact in a relatively short time. Costa Rica was unwilling to borrow from the Bank for several years, wanting to avoid further indebtedness, but a new project was recently approved by the Bank's Board (in June 2000). Perhaps the Bank's new instruments—the Learning and Innovation Loans and Adaptable Programming Loans—will fill an important new niche, helping fill a need for countries reluctant to take on a large debt burden (see Velozo and others 2000). Costa Rica's experience in developing successful policies and institutions for improved forest management needs to be disseminated to the Bank's other member countries.

None of the three forest-rich countries studied received any direct Bank assistance to the forest sector after 1991, although Brazil and Indonesia rank among the Bank's top seven borrowers. The Bank has applied forest sector conditionality to adjustment lending in Indonesia and Cameroon. China and India, by contrast, have been the two largest Bank borrowers, both overall and for direct loans for forest projects—together China and India have received 59 percent of the Bank's direct forest lending since 1991—but both overall and forest sector Bank lending to these countries is small relative to the countries' own overall investments. The Bank's experience in forest sector development in these two countries shows that it is possible to design programs that produce win-win outcomes, in both poverty alleviation and environmental management. Each country has a strong sense of ownership of its programs, although both lack the institutional and financial capacity to accelerate the implementation of their own programs. The Bank has had a relatively positive impact on the forest covers of both countries through a long-term partnership and mutual learning by doing. The economies of China and India have been more stable than those of forest-rich countries, and their reforms have been more gradual. Forest policy and institutional reforms have been brought about through the countries' own initiatives, which the Bank supported mainly through long-term project assistance.

The challenges to the Bank's involvement in the forest sectors of both China and India have increased. This is partly because controversies surrounding the treatment of ethnic minorities in both countries have increased the transaction costs of the Bank's continued involvement in the sector, while returns are in question. In India, both the Bank's country department and the finance ministry have sought more convincing evidence of the impact of forest projects and the fiscal sustainability of forest sector investments before proceeding with further Bank support in the sector. The future of the Bank's forest programs is even less certain in China, which has one of the most successful Bank-financed programs—even in production forestry, which yields higher financial returns to producers and greater economic returns to the country than does the conservation of protected areas. China's unique form of fiscal decentralization requires that all provinces, including those where there are forests, be fiscally responsible for the loan repayments to the Bank. In China, performance on all forest projects has exceeded appraisal estimates for tree plantings and yield growth. Reestimated rates of return on completed production forestry projects, for example, are 20 to 25 percent (see Rozelle and others 2000). The government of China questions the Bank's estimates, however, and estimates rates of return to be 10 to 13 percent, because of the high risks and variable performance associated with long-term forest investments and different species' varied performance in differing conditions.[6] Now that China has "graduated" from concessional IDA lending, its ability to borrow on IBRD terms for investments in the forest sector is in question. China's recent logging ban, which has affected many forested provinces, has reduced provincial incomes. The higher interest rate of IBRD loans to China has compounded the difficultly for poor provinces—and the poorest farm households within them—to mobilize the resources to undertake risky investments and to repay the Bank. This has drastically reduced the demand from China's poor provinces for Bank loans.

Market Forces and Deforestation

The 1991 Forest Strategy did not fully envisage how much the explosive growth in demand for forest and agricultural products in both domestic and international markets would exacerbate forest degradation, deforestation, and land conversion. Globalization, international trade liberalization, and lower tariffs have made forest-product imports from forest-rich countries cheaper in forest-poor countries, which has made domestic investment in trees less profitable. China (at $5.6 billion) and India (at $764 million) are among the largest importers of wood products, and growth in income, population, and urbanization has increased demand for forest products. According to government estimates, China's 1998 decision to ban logging in natural forests resulted in a decline in China's domestic timber supply of 16 million cubic meters a year. Imports—which were slight in the 1980s—have grown moderately in the past two decades, while exports have fallen. This trend is of regional and even global importance. Without countervailing investments in production forestry and tree plantations, China's rising demand for wood products could lead to defiance of the logging ban at home and continued exploitation of natural forests by local communities, or a surge in imports that exacerbates deforestation of natural forests in exporting countries, particularly in the East Asia Region. In India, although estimates vary, demand for industrial wood is expected to triple or quadruple from current levels in the next 25 years. Increased imports could also accelerate deforestation in countries that sell to India.

By increasing exports of both forest and agricultural products, forest-rich countries have increased pressure on forestlands. Sources of demand have varied. Indonesia has emerged as the largest exporter of tropical timber in the world. By contrast, nearly 86 percent of Brazil's vast timber is consumed internally, placing Brazil's per capita domestic consumption ahead of that of Western Europe. The development of plantations and of alternative sources of energy (such as natural gas piped through the recently opened Brazil-Bolivia pipeline) have been important in reducing the impact of Brazil's domestic consumption of wood products on its natural forests. But the environmental impacts of the substitution of gas for woodfuel are complex, and not necessarily benign. Cameroon's mostly foreign-owned timber companies have expanded exports to Europe, making Cameroon the largest exporter of tropical forest products in Africa. These domestic and global market considerations must become an integral part of the Bank's forest strategy.

Educating community forestry user group on how to collect fuelwood. Hetauda, Nepal. Photo courtesy of Still Pictures.

Land Conversion and Agriculture

Contrary to assumptions implicit in the Bank's 1991 Forest Strategy, no universal principles govern the influence of land conversion and agriculture on forestlands. How much agricultural intensification can reduce pressure on forests is highly location-specific and depends on such factor proportions as the rate and nature of technical change and transportation costs. The impact of agricultural productivity growth in land surplus countries such as Brazil has been to increase incentives for land conversion, a phenomenon reinforced by globalization and the liberalization of markets. But in densely populated countries such as China and India, agricultural intensification has helped relieve pressure on forestland. Land conversion to ex-

Deforestation: Using a computable general equilibrium (CGE) model to determine how the magnitude and impact of deforestation in the Brazilian Amazon are affected by changes in policy regimes and technology, Cattaneo (1999) shows that devaluation shifts agricultural production in favor of exportable products. How devaluation affects agricultural incentives in different regions depends on migration flows. If migration occurs only between rural areas, a 30 percent devaluation increases deforestation rates 5 percent. If urban labor is willing to migrate to the Amazon and farm, the deforestation rate increases 35 percent. San and others (1998) analyzed the short- and medium-term impacts of structural adjustment through devaluation on regional production, deforestation, factor markets, income distribution, and trade for the Sumatra region of Indonesia. The study found that devaluation encourages deforestation; exports of forest products as both final products and intermediate inputs for the wood processing industries increase. (See also Lele and others 2000c and Gautam and others 2000.) A recent piece of environmental sector work on Brazil argues that rainfall determines agricultural production's potential and may mute some of the effects of improved incentives from globalization (see Schneider and others 2000). Technical change, on the other hand, may expand agricultural options even in high-rainfall areas. These examples merely demonstrate the complexity and location-specificity of outcomes.

pand exports has become a particularly important issue in the three forest-rich countries—Brazil, Cameroon, and Indonesia—that have experienced irregular macroeconomic performance, huge pressures on the domestic budget, and trade imbalances. In the Amazon, land conversion for soybean production and livestock grazing is common. Returns to agricultural conversion are sizeable, partly because the forest sector is more heavily regulated than agriculture. Regulations without transparency and accountability, however, increase the potential for corruption and the cost of managing forestlands. Devaluation in the three forest-rich countries has increased incentives for all exports, but the agricultural exports have increased more than exports of forest products. The little evidence available on financial returns to managed forests suggests that without enforcement (evident in extensive illegal logging) and with virtually unrestricted agricultural sectors, there is little financial incentive for improved or low-impact logging to replace alternative uses of resources available to entrepreneurs.

Cameroon is atypical among the case study countries, but typical of the 80 Bank member countries that have experienced a decline in per capita income. Cameroon has regressed from being an IBRD borrower to being an IDA borrower, which makes economic growth an urgent necessity. Without improved forest management, Cameroon's rate of deforestation might well have been greater if its economy and exports had grown rapidly.

Forests and Poverty Alleviation

Our understanding of how forests and poverty interact would have been greater if the Bank's investments and ESW had characterized the forest-dependence of the poor with rigorous quantitative information.[7] Population pressure on the land, the incidence of poverty, and the role of forests in people's livelihoods vary in the six case study countries. In Indonesia the number of forest-dependent people is estimated to be between 1.5 million and 65 million, and in India between 1 million and 50 million. It is difficult to provide meaningful estimates because, unlike estimates of the incidence of poverty, reliable data are not available and definitions of forest dependency vary. It is important to distinguish between those who depend on forest products and services for survival and those whose livelihoods are improved by forest outputs. There is growing evidence that forest-based subsistence activities (low input/low output) do little to help people climb out of poverty and are rapidly abandoned once incomes begin to grow and better alternatives become available. And demand-driven forest activities that can be part of strategies of forest-dependent people for income growth and livelihood usually require certain levels of skill and capital, which not everyone has. Such issues, which have implications for the forests' role in poverty reduction strategies, and for program designs that focus on forest management, are not addressed adequately in the Bank's projects or ESW.

Moreover, improving the conditions for the poor generally and improving the conditions for indigenous communities can be conflicting goals. China and (until the East Asian economic crisis of 1997) Indonesia, for example, achieved impressively rapid economic growth and substantially decreased the percentage of people living below the poverty line. The records of both Brazil and India are less impressive on poverty reduction and improvements in social indicators. But in China the forested areas lie in mountainous regions that contain the majority of the officially designated poor provinces, including substantial minority populations. It appears, however, that China's recent graduation from IDA, the result of its economic growth, is likely to considerably reduce borrowing from the Bank for forest development in these poor provinces. India has a less impressive record on poverty reduction and improvements in social indicators than China. More than 300 million people live below the poverty line in

Returns to Low-Impact Logging: An extensive literature review on relative returns to conventional and low-impact logging (Pearce and others 1999) supports the findings of the low relative returns noted by the OED studies of Brazil, Cameroon, and Indonesia. According to Lele and others (2000c): "Protection of Brazil's Amazon forests beyond the short term requires an increase in the value of standing forest, an increase in the costs associated with unsustainable logging practices, and an increase in incentives for and profitability of sustainable (or improved) forest management. It must pay to keep trees and other forest products in the forest and improve management practices, and predatory exploitation of the timber must become unprofitable. In evaluating measures that might address these challenges, it is useful to distinguish between the processes taking place at and beyond the forest-agriculture frontier. At the frontier, agriculture, logging, and road-building create a mutually reinforcing system of forest conversion. Beyond the frontier, deeper in the forest, illegal logging of higher-value tree species threatens protected areas and the livelihoods of indigenous communities and extractivists."

India. Two-thirds of India's forest cover lies in tribal districts. The incidence of poverty among tribal people is high, and efforts to improve conditions have generated considerable conflict between tribal people and other poor people.

Development of the forest sector has helped the poor in China and India by:

- Creating wage employment in the planting and regenerating of forest areas.[8]
- Upgrading the skills needed to operate nurseries, graft fruit trees, and manage tree crops.
- Creating assets by either giving the poor rights to the trees on public lands or the rights to the public land itself to help meet subsistence needs for fuel, food, fiber, and other non-timber forest products. When villagers in China were allowed to plant the trees of their choice on previously owned public lands, the number of tree species increased from 4 to 16 and the emphasis shifted from timber to fruit trees, on which there are fewer harvesting and marketing restrictions (Rozelle and others 2000).
- Reducing overall risks and diversifying incomes for the poor by promoting tree planting as part of agricultural cropping systems.
- Creating social capital by increasing the collective ability of forest-dependent communities to plan, manage, grow, and equitably share common property resources. China is clearly far ahead of India in the development of local social capital for planning and implementation, mainly because China's "responsibility system" has virtually decentralized control of public lands to villages, communes, and individuals. But the recently imposed logging ban is introducing new restrictions on the rights of village communities.
- Many kinds of policy and institutional reform contribute to poverty alleviation. Production activities have received more attention than tenure arrangements, pricing, and markets for timber and non-timber forest products.

Institutional Issues

Institutionally the six countries vary from the decentralized democracies of Brazil, Costa Rica, and India, to the more state-controlled China. But China is more decentralized than it appears to be, having given a strong and active role to its provincial, county, and township governments. And community participation became more active in the 1980s when China introduced its "responsibility system," based on the devolution of land rights to households. The community forest management programs of China and India differ greatly from each other, partly because rights to forestland or its fruits devolved to communities less extensively in India than in China. China's recent centrally imposed logging ban, however, appears de facto to have taken away some of the property rights given to local communities. Cameroon and Indonesia have been slower to decentralize and encourage community participation. When the Indonesian government nationalized community lands (de facto) in 1967, concessions were awarded to the politically well-connected to maintain their political support (Gautam and others 2000). Devolution is now being introduced in Indonesia as part of political reforms.

States in the six countries differ greatly in governance capabilities and in capacity to manage the forest sector. Devolution, decentralization, the diverging interests of multiple stakeholders, and the likelihood of conflicts among them, all have implications for governance. There are also major differences in the forest sector administrations of the six countries. Costa Rica enjoys a highly diversified set of public institutions and committed NGOs and private associations that interact to address the many multisectoral functions and equally diffuse costs and benefits of the forest sector. Cameroon is the weakest in institutional capacity, followed by Indonesia and several states in the Brazilian Amazon. Even the forest producing states in western China are institutionally weak in several respects. In Brazil rents from forest exploitation are decentralized and broadly distributed at the local, municipal, state, and national levels and among different actors. In Indonesia such rents have been concentrated in the hands of a few who are close to political power, although that situation is changing rapidly. Because Brazil is also far richer than Indonesia in its forest resources, and its economy is less dependent on forests, unsustainable forest management poses fewer costs and huge benefits from Brazil's national perspective—although it entails huge costs from the global perspective.

Land Tenure

The 1991 forest paper stressed secure land tenure as a way to increase incentives to invest in trees and to reduce incentives for resource mining. But again, how much this proved true varied from place to place. How much tenure security affects land conversion to agriculture, and hence rates of deforestation, is a controversial issue in Brazil (Lele and others 2000c). Some argue that securing land title increases smallholders' access to credit, information, and extension services, thereby facilitating the clearing of land for agriculture. Others argue that farmers with no access to credit cut trees to finance their agricultural operations. Either way, the net effect in the short run tends to be deforestation. The merits of secure land rights for indigenous populations are more generally accepted. The government of Brazil has considerably accelerated the demarcation of indigenous reserves—some of it through Bank support—although many external threats lead to the de facto loss of land rights and forest cover.[9]

The issue in Cameroon and Indonesia is timber concessions. Until the mid-1980s the Bank considered customary tenure rights in Cameroon an impediment to the development of "unused" forest resources. It recommended overhauling land tenure legislation so that land expropriation in support of state and private development of industrial plantations could become operational (Nssah and Gockowski 1999). Now, however, Bank strategy favors improving Cameroon's legal and regulatory framework, although both the forest law and the implementation decree have failed to provide an adequate legal framework for planning land use and integrating forest conservation and production activities with agriculture. The prevailing land tenure regime assigned usufruct rights to anybody who cleared and cultivated land in the state-owned forests, which make up most of Cameroon's dense forest. This has probably encouraged deforestation. In Indonesia, however, the Bank recommended revenue-sharing with communities, but the government was uninterested in devolving rights to communities until democratization in 1999.

In China, the question of how land tenure has affected investment in forests and the protection of forest resources is part of a debate about whether the reforms of the 1980s that gave control of forestlands to farm households helped or hurt forest management. After successful reform to decollectivize in the agricultural sector in the early 1980s, leaders in the forest sector sought to further devolve control and increase incentives to households and forest users. The reforms did not go as far as those in agriculture in the early stages of implementation, but in the past decade several innovative programs have been launched in an effort to improve forest investments. The academic and policymaking communities disagree about the success of this movement in terms of forest sector outcomes in deciding the impact of the recent logging ban, but China's more equitable initial distribution of land and local power has ensured more socioeconomically equitable outcomes (Rozelle and others 2000).

Decentralization

Experience in Brazil, China, and India shows that, contrary to popular belief and some positive experience, decentralization does not always improve environmental management. It may even worsen it in the short and medium term, when local institutions are in their infancy and checks and balances are limited. Even if local capacity for environmental management increases substantially, it may not be an adequate substitute for the functions previously performed by the central government. The precarious finances of many states and municipalities is an increasing concern as countries decentralize. Logging and forest industries generate revenue and employment at several levels, making it possible for local governments to accept forest concessions. Politically powerful loggers and ranchers in Brazil (typically supported by municipal and even state assemblies) compete with small agents and indigenous people for control of forestland. National governments may be more concerned about national environmental objectives than their local counterparts and may pursue policies of conservation—such as China's introduction of a logging ban—which are seen as costly at the local level. But national economic and political interests may also be aligned with the state and local elite (as in Brazil, Cameroon, and Indonesia). Forest sector reform should address issues of interest to key stakeholders, such as employment, income generation, and government revenues. Most responses will involve investing in increased productivity for all forest products and services—investing, for example, in research, extension, and markets, areas the Bank has given less attention than it has to conservation issues.

The Bank's 1991 Forest Strategy and Forest-Poor Countries

Forest policies in China, Costa Rica, and India were developed independently of the Bank's strategy, but are consistent with it. India's 1988 forest policy has several elements in common with the Bank's strategy. Both emphasize the environmental role of forests and the subsistence requirements of forest-dependent populations. The Chinese government's logging ban brings China's policy more in line with the Bank's forest strategy. But the immense challenge faced by India and China in implementing their progressive policies can be better appreciated by looking at the case of smaller and richer Costa Rica.

The small amount of ESW the Bank has done in Costa Rica has had a positive impact on government policies. In its 1993 forest sector review, the Bank estimated that nearly two-thirds of the benefits of Costa Rica's forests are enjoyed globally, so the global community should compensate Costa Rica for conserving, managing, and planting forests. Few transfers have materialized, however, and the country is bearing the cost of its environmental policies largely on its own. To enable the Costa Rican government to finance such environmental activities as the protection of biodiversity, the review suggested improving the financial management of national parks. It also recommended deregulating harvesting in forest plantations and the import and export of forest products. Some believe that the Bank's policy advice to acquire private land for conservation instead of allowing private landowners to make a living from conservation activities may have been counterproductive. Critics also argue that the Bank's policy advice neglected the plantation sector, the development of small private forests, and post-harvest aspects of forest production—including helping Costa Rica attract more private investments. The governments of Brazil and China also stressed Bank Group financing for tree planting and post-harvest activities, but the Bank's financial sector policies since 1991 have brought about a sharp decline in the Bank's lending for such specific activities (World Bank 1998c).

The Bank's 1991 Forest Strategy and Forest-Rich Countries

The Bank's services have been less in demand in forest-rich countries than in forest-poor countries. The reasons for the Bank's small role and negligible impact have been different in each case. In Brazil the government has viewed the Bank's focus on saving forest cover in the Amazon as an unnecessary interference in Brazil's domestic affairs, although this attitude may be changing as Brazil's domestic environmental lobby gains voice.[10] In Indonesia the domestic plywood industry, working closely with the forest department, has resisted reform. In Cameroon the powerful foreign-owned industry working with the parliamentarians has resisted reform. But the Bank has continued trying to affect forest outcomes in Cameroon and Indonesia through adjustment-related policy conditionality and with increasingly active donor coordination and consultations with stakeholders.

The Bank's failure to have an impact in the three forest-rich countries seems to be the result of its "precautionary" strategy (focused on conservation, with no instruments to enable research and experimentation to improve forest management or deal with illegal activity) combined with weak implementation and the Bank's limited effort to nurture a national consensus for policy reform through country dialogue.[11] In addition, forest-rich countries—with their forest "surplus," low levels of royalties, and heavy (legal and illegal) logging—have had little incentive to use forests efficiently. It is important to increase the value of forests through stronger enforcement, to create new markets for environmental services, or to identify measures to fill the gap between the globally and nationally perceived optimal quantities of forest resources through measures such as transfer payments for environmental services. Without additional resources, budget-strapped governments find it difficult to afford stronger enforcement or payments for environmental services.

China's logging policy: The logging ban also has tremendous short- and medium-term social and economic costs. It has already reduced the timber supply and will probably affect the jobs and income of nearly 1.2 million people directly and another 1.2 million indirectly. In addition to a loss of revenues, the government is anticipating fiscal transfers of about $20 billion over 13 years from the center to the suffering provinces, and perhaps an increase in imports. This is happening at a time when the Bank's focus in China's forests is shifting from production to conservation and biodiversity and when China is shifting from IDA lending to more costly IBRD lending. Clearly, the forest lending program in China is coming under pressure.

The impact of the ban on Bank financing for commercial logging in tropical moist forests has been mixed. It discouraged the Bank from supporting experiments to address "improved and conservation-oriented forest sector management," but it also reduced the risk of Bank association with illegal, large-scale, and unsustainable logging. The ban has had strategic and symbolic value, but did not discourage wasteful practices in forest-rich countries, so it was largely irrelevant in containing rates of deforestation—a view shared by key borrowers and Bank staff (discussed in the next chapter). More important, the ecological risk aversion associated with the 1991 strategy had unintended effects: It discouraged the Bank from promoting changed attitudes and helping countries build internal capacity and, paradoxically, because the Bank was a nonplayer, it hindered Bank efforts to give conservation-oriented forest constituencies a voice in their country's internal decisionmaking. If the current ban on Bank financing of commercial logging in primary tropical moist forests is extended to the forests of Eastern Europe, where forests are already managed for multiple uses, it could jeopardize promising Bank efforts currently under way and could have a similarly chilling effect on the Bank Group's future ability to mobilize the private funding needed for continued responsible forest management.

In many forest-rich countries, local rent-seeking objectives in the forest sector may have restricted Bank involvement. The 1991 strategy also may have affected the demand for Bank services in an increasingly demand-driven Bank. Without substantial support to compensate for the fiscal and economic costs of conservation, governments have been reluctant to internationalize the forest issue, both because of the economic benefit they see in exploiting the forests for their countries' industrialization and modernization, and because of the close connection between resource exploitation and governance issues. In Brazil, as in Indonesia, economic crises and large-scale projects (such as Brazil's Rondonia Natural Resource Management Project) have brought these issues to the fore.

It is unclear how the Bank alone can stimulate the development of internal capacity, transfer skills, and mobilize the financial resources needed for sound forest policy without a broadly shared, cohesive, and consistent diagnosis of the problems by the international donor community and a broad range of national stakeholders. International objectives can differ from borrower-felt needs and priorities, as they have in Brazil. Until Mr. Wolfensohn's recent initiatives in the forest sector, the Bank and the donor community had not made enough effort to develop a common vision and understanding of borrower needs. The forest sector is well suited to the Comprehensive Development Framework (CDF), a "holistic" approach to development that emphasizes client ownership, the Bank's partnership with other development actors, better program design through Bank interaction with all stakeholders, and concrete results. Notwithstanding past and current differences between the Bank and some governments, especially in forest-rich countries, the question is, given the current strategy's ambitious goals and meager resources, can the Bank harness such emerging opportunities as presented by increased environmental awareness, active domestic stakeholders, and active environmental institutions (especially NGOs) to improve forest management. This question will be addressed in the next chapter.

Global Trends and Changes Affecting Forest Policy

How current is the World Bank's forest strategy? Some of the technological, institutional, economic, and policy changes that have affected forests since 1991 must be considered when the Bank's forest strategy is updated. But some trends are difficult to document because definitions used for forest data often differ across countries and many national and international statistics on forest cover are unreliable. Production data are suspect because of substantial illegal logging. Data inconsistencies within countries over time impede understanding of the extent, sources, and causes of loss and degradation of forest cover, and efforts must be made to improve the quality of forest data (Annex B).

The formulators of the updated strategy will certainly emphasize the building of partnerships, but they will also have to take into account that past efforts at coordination, including Food and Agriculture Organization (FAO) supported National Forest Programs (NFPs) and Bank-encouraged Environmental Action Plans, have often not worked. The NFPs, which succeeded the Tropical Forest Action Plans in 1985, were supported by the FAO, the World Resources Institute (WRI), and the Bank. In due course, however, the Bank and WRI dropped their support for these programs because they were "top-down" and narrow in scope. The NFPs continued as FAO initiatives carried out by developing country governments with the support of several bilateral donors and the European Union. The Bank, meanwhile, encouraged countries to develop Environmental Action Plans, which in many ways resemble NFPs, but with a broader scope and often variable treatment of issues in the forest sector. The Bank, FAO, and the donor community have not streamlined or coordinated their activities to bring about essential policy and institutional reforms that reflect the concerns of all stakeholders, especially the poor. What is clearly needed is a better-coordinated country- and client-driven approach. The collective international effort being undertaken in Cambodia, led by the Bank with the active participation of the FAO, U.N. Development Program, and other donors is an example of successful coordination, albeit at an early stage.

Changes in the Forest Sector

Global wood production and consumption is likely to increase about 26 percent between 1996 and 2010 (Whiteman 1999), with increases throughout the developing world, but especially in the Europe and Central Asia and East Asia and Pacific Regions (table 4.1).

Production and consumption are in close balance within Regions, but there is substantial intraregional trade. Demand from China, India, and Japan, for example, has been met largely by Southeast Asian countries (Dauvergne 1997). More than a fifth of the global consumption of industrial wood is expected to be concentrated in Brazil, China, Indonesia, and the Russian Federation.

To contain deforestation and the degradation of forest resources, countries need substitutes for products originating from natural forests. Investments in alternative sources of wood—ranging from small-scale tree planting to forest plantations on degraded lands—are becoming important. Plantation forests now offer great promise for wood production.[1] Improved genotypes and recent advances in silvicultural technologies have led to spectacular yields. Annual growth rates of 20 to 30 cubic meters per hectare a year are common on research stations in tropical areas and are being realized on many private lands. Industrial wood from one hectare of a plantation can easily substitute for 5 to 20 hectares of natural forests (Binkley 1999). Plantations not only relieve pressures on natural forests but also offer reliable supply, uniformity of product, and competitive economics. They could potentially satisfy most of the global demand for forest products (LEEC 1993). Moreover, plantations provide environmental benefits, reducing soil erosion, protecting watersheds, conserving biodiversity, and sequestering carbon. Large or small tree planting operations, if conducted in an economically, environmentally, and socially responsible manner, could provide the poor with tremendous employment opportunities, either as tree owners or as workers on plantations and in post-harvest operations. They can also reduce pressure on natural forests.

TABLE 4.1. EXPECTED INCREASES IN INDUSTRIAL WOOD PRODUCTION AND CONSUMPTION (1996–2010, MILLION CUBIC METERS PER YEAR)

	1996	2010
Region	**Production** *(consumption)*	**Production** *(consumption)*
AFR	66 (59)	81 (72)
EAP	304 (339)	421 (466)
ECA	380 (380)	547 (556)
LCR	141 (131)	164 (143)
MNA	7 (8)	10 (11)
SAR	31 (31)	57 (58)
Other	560 (546)	593 (574)
World	1,490 (1,493)	1,872 (1,881)

Source: Whiteman, Brown, and Bull 1999.

Only a decade ago, intensively managed forests supplied only 7 percent of industrial wood; now they supply about 26 percent and are expected to supply about 40 percent by the year 2040. More than half of wood production in industrial countries is now from managed forests and plantations. Productivity is already high in plantation forestry in Brazil, which has a progressive private sector and a strong research system that is funded partly by the private sector. But this is not typical in other developing countries, and small-scale planters need investment help in all developing countries. Real interest rates tend to be high, making it next to impossible for small farmers to obtain credit. In Brazil and China private companies have made surprisingly similar suggestions: Give Bank/IFC loans to processing enterprises that enter into partnership with small farmers for plantation forestry, while government provides research and extension services with appropriate environmental and social impact assessments in place.[2]

Investments are being made in industrial plantations in countries as diverse as Argentina, Brazil, Chile, Indonesia, New Zealand, South Africa, and Uruguay. Discussions of the CEO Forum (box 4.1) suggest that prospects for potential investments in developing countries are strong if a favorable investment climate is ensured (Annex G). The Bank Group institutions—IBRD, IFC, and MIGA—need to work together to encourage more private capital—on a scale appropriate to the specific country circumstances—to invest in socially and environmentally responsible plantation forestry. The Bank's experience in Brazil and China offers some important lessons about undertaking socially and environmentally sustainable plantation forestry. The government of the state of Minas Gerais in Brazil is so convinced of the merit of the credit extended under a previous World Bank loan to help small farmers undertake forest plantations that it has converted it into a commercially operated revolving fund. In several developing countries, particularly in the tropics, millions of hectares of degraded lands are suitable for tree planting.

Investments in plantation forestry in these countries—combined with an increased commitment to bringing natural forestland under protection, while ensuring productive livelihoods for forest-dependent communities—can meet urban and export demand, create livelihoods, and substantially reduce continued pressures on the world's natural forests.

Tree planting operations will succeed in replacing wood from poorly managed and unsustainable operations in natural forests only if tree planting is economically competitive with products originating from natural forests. It is crucial to improve governance and to eliminate the perverse incentives, market distortions, and constraints on valuation currently associated with natural forest operations. To develop economically and environmentally competitive tree planting operations, it is also necessary to provide clear and secure property rights, access to financing, and payment for environmental services. In countries such as Costa Rica, where legislation has supported such initiatives, tree planting has successfully changed the extent of forest cover; in countries without such incentives, tree planting and investment in plantations have plummeted.

Developing countries have greatly expanded their protected areas in recent years but have not improved management in existing *or* new protected areas. Environmentalists welcome the extension of protected areas as an indication of the countries' commitment to conservation. Both the Bank and GEF have supported such expansion through project assistance, even before

BOX 4.1. THREE NEW WORLD BANK INITIATIVES

The World Bank has three new initiatives to improve the conservation and sustainable management of the world's forests.

The World Bank/WWF Alliance (formally the Alliance for Forest Conservation and Sustainable Use). In April 1998, the World Bank and WWF entered into an alliance to work with governments, the private sector, and civil society to reduce the loss and degradation of all types of forest worldwide. The alliance partners propose to work together to support countries to achieve the following targets by the year 2005:

- Establish an additional 50 million hectares of protected forest area, and bring a comparable area of existing, but poorly managed, reserves under effective protection.
- Bring 200 million hectares of the world's production forests under independently certified sustainable management.

The CEO Forum. This ad hoc group—which includes 31 representatives from the World Bank Group, the private sector, civil society, and governments—was assembled in 1997 to consider and discuss global forest-related issues, especially options for reducing barriers to sustainable forest management by promoting responsible investments in forest production and management. Participants have agreed that the discussions should continue as long as they contribute to an understanding of the issues and to practical initiatives associated with the preservation and better use of the world's forests.

Prototype Carbon Fund. This initiative is a response to opportunities offered by the mechanisms of the Kyoto Protocol for mobilizing new and additional public and private resources and technology flows to member countries. Governments that signed the International Convention on Climate Change have still not decided whether to allow payments for carbon offsets involving forest creation and conservation, so the Fund is currently designed to operate only in the transition economies of Eastern Europe. Learning from that experience can be put into practice in developing countries if an agreement on payment for carbon offsets is reached. The Fund's objective is to explore how market-based mechanisms could help reduce the global concentrations of greenhouse gases and contribute to the sustainable development of the Bank's borrowers. A major emphasis would be placed on renewable energy technology. The Fund will support project-based activities that reduce greenhouse gas emissions and enhance carbon sequestration, but the Fund will not be involved in "emissions trading" activities.

the World Bank/WFF Alliance. However, the assistance did not involve analysis of the threats to protected areas from economywide and global policies, did not support the enforcement of existing laws and regulations, did not help build local institutional human and financial capacity, and did not generate the sustained domestic and international financial resources needed to support better management of existing protected areas, including better treatment of the people living in them. Besides, a highly restrictive definition of primary forests adopted in the 1991 strategy essentially banishes human presence or activity and is not conducive to conservation. It would require new definitions of primary and secondary forests more attuned to the realities in developing countries on the ground, improved fiscal policies, with transfers from polluting to greening industries across sectors, the introduction of fees for parks, greater revenue-sharing with local communities to increase their stake in protection, more mobilization of private capital, stronger links with local NGOs and research institutions, and the exchange of information about successful experiences within and across Regions. A wealth of experience exists in the Bank and GEF portfolios, especially in the Latin America Region. Tapping that experience calls for the multisectoral approach the 1991 Forest Strategy envisaged. The World Bank/WFF Alliance could address such fundamental issues as incentives and financial resources in its quest to increase protected and sustainably managed forest areas. The World Bank Institute's link with operations should be strengthened to realize this potential.

Changes in the Global Economic Environment

Patterns of global economic growth and income distribution are far different today than when the Bank's forest strategy was published. Per capita incomes in more than 80 countries are lower today than they were a decade ago. Only 40 countries have been able to maintain an average per capita income growth rate of 3 percent or more during the 1990s. The richest 20 percent of the 174 nations reviewed by the U.N. Development Program for its 1999 *Human Development Report* have enjoyed much of the recent global boom. Those countries produce 86 percent of global gross domestic product, whereas the poorest 20 percent produce 1 percent. Global exports are similarly distributed even though total exports are substantially larger today than 15 years ago. Of the 34 countries with the lowest human development indicators (based on income, education, and life expectancy), 29 are in Africa (UN 1999a).

Slow or declining growth in developing countries stems in part from the slow growth in demand for commodities. Real commodity prices are the lowest they have been in several decades. Many low-income countries that rely on exports of natural-resource-based products to earn foreign exchange must ship even greater quantities of forest, agricultural, and mining products to earn the same revenue. After accounting for the rapid loss of natural capital, adjustments to economic crises by developing countries undermine the sustainability of their long-term export and growth performance. But increased export growth is crucial to macroeconomic stabilization in the short and medium term and critical for a return to a sustained growth path for countries in economic crisis. These issues are seldom addressed in adjustment programs or by the global community, which has higher expectations for natural forest maintenance in developing countries than the development needs of those countries dictate.

Increasing population growth in developing countries raises demand for agricultural production and exacerbates pressures on forests. Currently about 1.4 billion hectares of cropland meet global food requirements. The FAO estimates that another 1 billion hectares of grasslands and 800 million hectares of the world's remaining forests and open woodlands could potentially be converted to crop production. The extent of forest conversion will depend on the level of continued agricultural productivity growth, although its effects on forest cover will be different among countries.

Changes in the International Institutional Setting

Many international agreements reached in recent years have profound implications for forest policies and programs in developing countries (box 4.2). There is increasing international cooperation on forest issues, but no global legal instrument deals specifically with the protection and management of forests. And except for the GEF, no new funding mechanism has been initiated to implement international initiatives. The "Rio forest principles" adopted at the United Nations Convention on Environment and Development in 1992 are a significant step in that direction, but a shared operational vision of sustainable development still needs to be developed. The Rio forest principles, based on concepts supported by a broad international consensus, cover issues ranging from sustainable development and biodiversity to trade in forest products and international cooperation. These principles recognize that countries have sovereign rights over their forest resources and urges countries to incorporate the principles in their national forest policy and legal frameworks. The principles encourage governments to promote the development, implementation, and planning of national forest policies and to provide opportunities for the participation in that process of interested parties, including local communities and indigenous peoples, industries, labor, nongovernmental organizations and individuals, forest dwellers, and women. But OED country studies have noted a major gap between principle and practice in countries that are signatories to the agreements. What should be the Bank's role, if any, in ensuing adherence to international conventions through its operational mechanisms? This question has received little attention in Bank operations (Gautam and others 2000), but given the Bank's unique professional and financial resources and its economywide experience in borrowing countries, it could play a major role in translating a global vision into action.

BOX 4.2. INTERNATIONAL INSTITUTIONS AND AGREEMENTS RELEVANT TO GLOBAL FORESTS

The **Intergovernmental Panel on Forests** (IPF) was established by the U.N. Commission on Sustainable Development (CSD) to continue the intergovernmental dialogue on forest policy at its third session in April 1995; to implement forest-related decisions of the U.N. Conference on Environment and Development, nationally and internationally; and to promote international cooperation in financial assistance and scientific and trade issues, among other concerns.

The **Intergovernmental Forum on Forests** (IFF) was established in July 1997 as an ad hoc, open-ended forum under the CSD. IFF has a mandate to promote and facilitate the implementation of IPF proposals.

The **World Trade Organization agreement** on trade liberalization has been at the center of the Bank's structural adjustment programs (now nearly one-third of Bank lending). Removing agricultural subsidies in industrial countries and continued liberalization of agricultural trade in developing countries would have complex effects on forest conversion.

The **Convention on Biological Diversity** promotes the establishment of national strategies for the sustainable conservation of biological diversity and their integration into sectoral and cross-sectoral plans and policies. Countries agree to undertake programs to identify and monitor biodiversity (including important ecosystems and habitats), to establish a system of protected areas, and to develop guidelines for the selection of protected areas.

The **Global Environment Facility** (GEF) is a mechanism for financing actions to address the loss of biodiversity, climate change, the degradation of international waters, and ozone depletion.

GEF-sponsored biodiversity programs in developing countries have been restricted largely to conservation in protected areas. The GEF is only now looking into issues related to the sustainable use of forest ecosystems. Some have argued that biodiversity is directly at odds with improved forest management in production forests, an issue that is also germane to discussions of the Bank's 1991 Forest Strategy.

The **Framework Convention on Climate Change** (the Kyoto Convention) offers promise for forests in their role as carbon sinks. Carbon trading may raise financial resources for investment in reforestation, afforestation, forest management, and conservation. Some people are concerned that too much attention to carbon trading will divert attention from other forest functions, such as supporting biodiversity and indigenous populations.

The **Agreement on Trade-Related Aspects of Intellectual Property Rights** may affect research in forest-related biotechnology and could have a profound impact on the rights to forest genetic material, competitiveness in international markets, and the livelihoods of forest-dwelling people.

The **Convention on Combating Desertification** is relevant to forest-poor countries, principally in Africa and the Middle East.

New Thinking and Policy Experiments

The idea behind the relatively new practice of forest certification is to establish a market-based incentive framework for improving forest management practices and to provide markets for products generated from sustainably managed areas.

The World Bank/WWF Alliance promotes improved management of 200 million hectares through certification. Certification activities are dominated by large firms located mainly in industrial countries with well-established forest policies and institutional frameworks. Demand for certified forest products and their supply are concentrated in Europe and North America. Certification could be a powerful instrument for improving forest management in developing countries, diversifying the sources of accreditation of forest management (currently monopolized by governments), improving forest management technologies and practices, stimulating capacity building for sound environmental management among consumers and producers of forest products, and generally increasing environmental awareness. To succeed, however, certification must operate in countries that use market-based instruments to internalize environmental and social externalities. For certification to be effective, it must have government support and countries must have growing demand for certified products, national standards for improved forest management (standards that are compatible with international standards where exports are involved), and an enforcement mechanism to ensure that the standards are met and the results relayed to major stakeholders in a transparent, credible manner. The certification process must also be set up in a way that enables rural communities to participate actively. In countries with weak institutions, certification could encourage corruption. Although it has a better prospect in channeling export demand to certified products through value addition than to influence internal demand in developing countries, the practice could also become a nontariff barrier for exports of products from developing and Eastern European countries, if not applied across the board to timber produced in all countries. But the certification issue is leading to far greater debate in forest-rich developing countries about improving forest management than the 1991 strategy could have envisaged (Lele and others 2000).

Several international forest product certification schemes—including the Forest Stewardship Council and Pan European Forest Certification—are developing rapidly, with the help of organized buyer groups, NGOs, and other institutions. National and regional certification systems are also emerging in forest-rich countries such as Brazil, Canada, Finland, Ghana, Indonesia, Malaysia, Norway, and the countries of the European Union.

The debate on sustainable forest management has advanced considerably since the Bank's strategy was published, but there is no global consensus on how to define "sustainably managed forests," partly because of the tremendous diversity of conditions in which forests exist. Each certification scheme has its own criteria and indicators. Recently the Center for International Forest Research sent teams of local and international experts to eight countries (Austria, Brazil, Cameroon, Côte d'Ivoire, Gabon, Germany, Indonesia, and the United States) to evaluate the validity and usefulness of the criteria and indicators used by various groups. The teams found general agreement on the main components of sustainability, including 6 basic principles and about 25 criteria related to policy, ecology, production, and social factors. Indicators and criteria tend to be site-specific, so that evaluators will probably need to adapt them to specific site characteristics. The Bank should—without associating itself with any particular certification scheme—support work to establish criteria and indicators around a common set of principles and to develop broad institutional capacity and human capital in borrowing countries.[3]

The Bank's recently approved Prototype Carbon Fund is an example of the type of experimentation the Bank should be promoting to remain at the forefront on environmental issues (see box 4.1). This program, which currently operates in transition economies, could make a positive contribution to the global environment (once international agreements are in place) without attracting the controversy that has threatened other similar initiatives. The Fund's target size would be $100–120 million, with required contributions of $10 million for public sector participants and $5 million for private sector participants.

Changes in Strategy Implementation

Since the 1991 strategy was formulated, the Bank has changed in ways that could have implications for the way it participates in global forums and the operational support of the forest sectors of developing countries. With staff levels declining, the Bank's participation in forest-related global forums has been

Clearcut. Washington State, U.S.A. Photo courtesy of Still Pictures.

haphazard. The Bank has participated in activities related to the Kyoto Protocol and some important new initiatives related to climate change, but its participation in the Intergovernmental Panel on Forests and Intergovernmental Forum on Forests, for example, was limited and unpredictable until two years ago. If the Bank is to be proactive in mobilizing enough concessional or grant resources to make a difference to global forest outcomes, it will need to follow a coherent approach to involvement in global processes that affect the forest sector. It will also need to support objective professional analysis of the extent and causes of deforestation in developing countries, with the active participation of stakeholders in those countries.

Many Bank operations staff, including department heads, have been relocated from headquarters to their client countries to strengthen Bank-client relationships and to respond to client-driven demand. The Bank's own decentralization has led to the adoption of a more open, consultative stance toward stakeholders in policy dialogue, in the preparation of CASs, and in some lending operations. These developments may not have been consistent with the expectations of containing deforestation to meet global objectives as espoused by the 1991 Forest Strategy.

The Bank's new emphasis on knowledge is a response to technological change and the information revolution. Networks of technical specialists in the Bank are now expected to facilitate knowledge dissemination; enhance product quality; and provide faster, more effective responses to constraints identified during project operations. The electronic revolution has enabled the Bank to communicate with professionals and clients throughout the world, particularly in borrowing countries, in a way that was not possible only a few years ago. This new orientation is critical to the CDF and should enhance the Bank's work on forests, which is often multisectoral and depends heavily on the transparency and accountability associated with improved governance.[4] However, the tremendous decline in the number of forest sector professionals may limit the Bank's effectiveness at seeing forest conservation in the context of broader global, technological, and market considerations.

Three Perspectives on Forest Strategy

This review, in addition to drawing on extensive informal consultations with stakeholders, professionals, and the NGO community in borrowing countries, surveyed several groups through interviews, focus groups, and formal surveys, including Bank staff who work on forest issues, participants in the CEO Forum, and members of the World Bank/WWF Alliance. The formal survey results support many of the conclusions OED reached through its Regional portfolio analysis, country case studies, and focus group discussions about the Bank's unrealized potential to play a more active global role.

Bank staff survey. The Bank staff surveyed agree with the overall thrust of the 1991 strategy but do not believe the Bank has effectively helped reduce rates of deforestation. They think forest issues are not well integrated with the Bank's broader mission of poverty alleviation and economically and socially sustainable development, that more attention should be paid to the forest sector in CASs and ESW. The staff believe the policy of banning Bank financing of commercial logging in primary tropical moist forests is irrelevant and has not affected the rate of deforestation in client countries. Many believe that making the Bank policy more flexible than OP 4.36 would better address key forest management issues associated with logging in primary tropical moist forests. They also believe Bank performance in the forest sector should be strengthened to promote greater protection of natural forests, institutional reforms, multisectoral approaches to forest development, and the planting of new trees.

Staff say that forest-related work within ESSD would be more effective if forest sector leadership, operational support, and resource flows were consolidated. They believe that to convert ideas into activities, the innovative strengths of the Environment Department needs to be combined with greater operational support than the Rural Development Department has traditionally provided. ESSD has recently reorganized the forest sector team, first with two co-managers—one each from the Rural Development and Environment departments—but many challenges remain. Every vice-president of ESSD has stressed the management challenges of ensuring central leadership in the forest sector, because the sector straddles many disciplines and Bank Networks. Staff believe country managers are more likely than staff in other sectors to consider involvement in the forest sector as entailing higher transaction costs and lower payoffs than other activities, which inhibits responsiveness to the challenges of improving forest management. Reasons given for these perceptions include internal constraints such as inadequate resources for ESW and project preparation and supervision; the complexity of project design (which for forest activities is inherently multidisciplinary and multisectoral); flaws in the management structure of the Bank; an inadequate skill mix; and internal technical and budgetary constraints. External constraints include corruption in implementing agencies, inadequate appreciation of key issues by policymakers in borrowing countries, insufficient implementation capacity, less-than-influential forest and environment ministries, and the controversial nature of forest-related policies. Staff believe the Bank could become a global leader in such forest-related issues as climate change, biodiversity conservation, natural resource management, carbon sequestration, and the Clean Development Mechanism. With a clear strategy and the right internal incentives and resources, the work of both the GEF and the IFC could increase.

CEO Forum survey. The CEO Forum has increased awareness of the Bank's 1991 Forest Strategy (and its safeguard policy content), especially in the private sector. The open discussion of forest issues between private sector and NGO representatives has revealed many areas in which their views are similar, and some participants believe that a number of their views have converged. Several private sector participants have asked for a broader-based approach to promoting improved forest management and reforms and more attention to the economic aspects of sustainable development. Others think favoring conservation over sustainable forestry will erode the Bank's credibility. Most agreed that the policy focus should be broadened to include all forest types, not just tropical moist forests. Members of the forum acknowledged the benefits from certification, but like the IFC voiced concern about its costs and benefits, end-users' willingness to pay a premium for certified products, the need to agree on guidelines, problems of third-party monitoring, and the possibility that certification could become a barrier to trade. Some forum members believe the Bank is well-positioned to address issues of climate change, biodiversity conservation, and resource management, although some NGOs disagree. A majority of forum members felt that the role of the GEF and the IFC should either increase or stay the same, but almost half of the private sector respondents did not know enough about the GEF's activities to respond.

World Bank/WWF Alliance survey. Respondents believe that the World Bank/WWF Alliance strengthens the environmental expertise, and uses the comparative advantages, of both the Bank and the WWF. The alliance's main potential benefits are increased funding for conservation activities and increased attention to good practices in improved forest management. Its drawbacks include the "top-down" approach it takes, its limited financial resources, and the time of Bank staff for alliance activities without commensurate budget support.[5] Most WWF staff believe the target of establishing 50 million hectares of new forest-protected areas to be realistic; they are much less certain about whether 200 million hectares can be placed under independently certified management by 2005.

Bank staff consider both goals to be unrealistic under current internal and external constraints. Alliance members believe fuller integration of the alliance agenda with the Bank's agenda to be essential. They stress the importance of developing a longer-term vision with clearer regional targets and better communication and reporting. Either more resources should be allocated to the alliance or its clients should be fully informed about budget constraints within the Bank and fiscal constraints in the countries, so that expectations become more realistic. Both WWF and Bank staff consider decentralization, realistic targets, and increased stakeholder participation important ways to make the alliance more effective.

5 Conclusions

This review of Bank activities and changing Bank and international trends supports two main findings. First, the Bank has implemented its 1991 Forest Strategy but with significant shortcomings. Second, that implementation has had only a modest impact on the strategy's two central objectives: slowing deforestation in tropical moist forests and planting new trees. The 1991 strategy made Bank activities more conservation-oriented, but it did not provide for an effective means of implementation.

Despite the Bank's limited presence in the forest sector—less than 2 percent of its total lending—it could be a highly influential global actor. But the task is challenging: the Bank's goals are ambitious, yet it has allocated very limited resources to the sector. Furthermore, stakeholders' expectations are diverse. For example, neither the high conservation expectations of some NGOs nor the contrasting perspectives of the private sector are necessarily shared by the Bank's borrowing governments. This review concludes that the Bank needs to adopt a broad-based strategy that includes the views of all stakeholders, more proactively pursuing the twin objectives of conservation and development in a financially and fiscally sustainable and socially equitable manner that stimulates genuine borrower demand and simultaneously achieves the Bank's central mission of poverty alleviation.

Strategy Implementation

The Bank has only partially implemented the 1991 Forest Strategy, mainly by increasing the number of forest components in its environmental lending—in projects that aim to conserve forests, enhance forest-related services, and improve the livelihoods of the people living in and on the margins of forests. The projects' diverse approaches are consistent with the strategy's focus on policy and institutional reform, improvements in forest inventories, forest management and regeneration plans, and participatory approaches. With the notable exception of China, however, tree planting and non-timber forest products and services have received much less support in Bank lending. In China the Bank has successfully supported the expansion of tree cover by nearly 3.9 million hectares through tree planting on different scales and under a wide array of tenurial arrangements—from involving poor minority households on individual, community, and public lands, watersheds, and shelterbelts to large-scale public plantations—with significant national and global environmental benefits. However, marketing of timber and non-timber forest products are issues even there.

Outside the confines of the forest sector, Bank activities have not paid enough attention to certain critical factors—often external to the forest sector—that affect implementation of the policy's central objectives. Integration of the principles of the forest strategy into the Bank's CASs, macroeconomic policy advice, and adjustment lending has been limited, and adjustment lending has increased substantially since 1991. Even where those principles have been integrated, the link between these macro efforts and Bank operations has been weak, a characterization that is not unique to the forest sector. The 1991 strategy did not anticipate the powerful effect of macroeconomic policies and globalization. And the multisectoral approach and international cooperation emphasized in the strategy have not been adequately pursued.

The strategy, along with independently evolving Bank internal policies and processes, such as environmental impact assessments and other safeguards, has prevented operations in sectors such as infrastructure from contributing to deforestation. This is a substantial achievement. But sectoral analyses have been spotty in addressing the impact on forests of non-forest-sector policies. And although safeguards have been applied at appraisal in non-forest-sector lending operations, monitoring and evaluation of the impacts of those safeguards has been inadequate.

Direct forest lending (for mainstream activities of forest ministries and departments) has stagnated. Nearly two-thirds of it is concentrated in China and India, where it is at risk of decline—in China, because of the shift in the Bank's lending terms; in India because of competing demands for borrowed resources. Forest lending has plunged in Africa, where the need for forest assistance is greatest and where the poor are overwhelmingly dependent on forest products and services. Forest sector lending also has not been sufficiently integrated into the Bank's agricultural, rural development, or poverty alleviation strategies, although some of the world's poorest people rely on forest products and services for their livelihood. Prominent among the forest-dependent poor are women, but gender considerations have also received too little attention in the Bank's policy implementation. Given the complex cultural challenges of getting women involved in poverty-oriented forest projects, the Bank should make extra efforts to consult and engage nationals from borrowing countries knowledgeable in these issues.

These criticisms notwithstanding, some noteworthy changes are under way in sectoral analyses and adjustment lending in the East Asia and Pacific Region, and in increasingly sophisticated participatory approaches being used in the East Asia, South Asia, and Latin America and Caribbean Regions. More participatory and phased approaches, including learning by doing, are being incorporated into project design in the Eastern Europe and Central Asia Region. These are important new pilot approaches to implementation, but it is too early to assess outcomes, effectiveness, and sustainability. The Bank's challenge will be to integrate the substance of forest sector issues into CASs, to better link the CASs to operations, and to provide the right incentives for scaling up the experiments with positive outcomes to achieve more impact.

In addition to its regular operations, since 1991 the Bank has launched new initiatives in response to changing global circumstances and increasing concerns about deforestation. Initiatives such as the World Bank/WWF Alliance and the CEO Forum seek to increase the dialogue among stakeholders, to develop a consensus on such contentious issues as the role of certification and how to define "protected areas" and "sustainable forest management." However, the new initiatives have not been fully integrated into the country or forest sector strategies and their sustainability is not assured (see Annex G).

Effectiveness

The effectiveness of the 1991 strategy has been modest with respect to achieving its two main objectives, and the sustainability of its impact is uncertain. The rate at which tropical moist forests are disappearing has not been reduced through Bank efforts, and tree planting with Bank financing has occurred on too small a scale to make a significant impact on global forest cover. The strategy has kept the Bank from getting involved in policies and operations that harm forests, thereby absolving it of "guilt by association." But the world's forests, especially the tropical forests, continue to deteriorate and the strategy overlooked temperate, boreal, and tropical dry forests, which are also socially and environmentally important.

There are three related reasons for this mixed performance; they have to do with how the strategy and subsequent policy were conceived. First, because the rapidly changing forces that were reducing forest cover and forest quality were not correctly diagnosed, the remedies suggested were either inadequate or misdirected. OP 4.36, the unclear "do no harm" policy, tried to give Bank staff more flexibility in addressing the challenges of forest management, but responses to staff surveys indicate that it did not succeed. Second, the consultative process used to develop the strategy was not inclusive enough. It left out key borrowing country and private sector perspectives, as well as those of Bank staff and managers, many of whom developed no sense of ownership of either the strategy or the policy. Some Bank client countries were unaware of the forest strategy until this OED review. Moreover, the strategy led to a significant loss of the Bank's internal capacity, as a number of forest specialists left the Bank in the years after the strategy was announced. Finally, the rapid changes since 1991, both globally and in the Bank, have made the strategy less relevant.

The strategy's limitations. The strategy's focus was too narrow—20 countries with tropical moist forests—and neglected other biodiversity-rich types of forest. And in assuming that poor people were the main cause of deforestation in those tropical moist forests, it failed to consider the role of other actors. The strategy overlooked fundamental governance issues that affect forest sector development and did not consider forests integral to the Bank's poverty alleviation mission. It focused largely on economic solutions, such as the length and price of concessions (as incentives for conservation), and on private property rights. In reality, a wide range of often complex property arrangements are encountered in forest sector development. The strategy diagnosed the problem of externalities (the divergence of global, national, and local costs and benefits) as a factor in deforestation but, except for the small GEF grant program, it did not call for mobilizing enough resources to meet the resource gap for conserving forests of global value. It assumed that governments would borrow funds on IBRD or IDA terms to achieve global (or even national) objectives for forest conservation, although they had other, more pressing, priorities.

Disincentives to implementation. Within the Bank, insufficient staff ownership of the strategy, shortcomings in human and financial resources, and a lack of incentives have been obstacles to the strategy's full implementation. Country and task managers and client governments perceive Bank involvement in the forest sector as entailing higher transaction costs and reputational risks than involvement in other poverty-alleviating sectors. Contributing to this perception are the ambiguous nature of OP 4.36's conditions for Bank involvement in sustained-yield logging, the associated ban on Bank financing of commercial logging in primary tropical moist forests, and the requirement that projects may be pursued only where there has already been broad sectoral reform. Even though the ban was confined to primary tropical moist forests, the controversy surrounding *any* logging in *any* natural forest has led to the perception of reputational risk for Bank involvement in other forests, including secondary natural forests in the tropics. Even in client countries heavily committed to conservation, the costs of dealing with the Bank in the forest sector are widely viewed as disproportionate to the potential benefits.

Decreasing relevance. The Bank's client countries have viewed the Bank's limited funds as being better suited to competing uses with quicker payoffs, even for the objective of poverty alleviation (girls' primary education, rural electricity, drinking water supply). The returns on long-gestating forest investments are often not high or quick enough, but the risks tend to be too high, relative to other possible uses of those funds, to stimulate borrower demand for Bank involvement or investments. This has been true not only for tree planting involving poor households in China—where demand for Bank funds has plummeted as interest rates have increased—but also for medium- and large-scale plantations in Brazil's private sector. The 1991 strategy did not adequately acknowledge either the long-term nature (and the short-term costs) of forest-related benefits or their implications for financing. Whether as part of forest strategy or of financial sector policy, access to finance for tree planting on any scale has diminished. Similarly, in retrospect, the strategy understated the power of domestic and international market forces to strengthen the incentives to cut trees or to place land under alternative uses, increasing deforestation. With or without Bank-financed structural adjustment, borrowing countries have often liberalized their economies in the context of globalization to take advantage of market opportunities that would help them maintain or resume growth and alleviate poverty. Liberalization has increased incentives for deforestation in many of the Bank's client countries, often with adverse impacts on forests and the people who depend on them.

Drawing criticism: The strategy's definition of "primary forest" and the application of Bank safeguards, particularly those on indigenous people and resettlement, have added to external criticism, including the threat of inspection panels. These controversies have been significant in forest-rich countries, which have wished to use their substantial natural forests for financing economic development, but they have also extended to other types of forests and projects.

Reconsidering the Bank's Forest Strategy

Efforts to promote forest sector objectives, especially the conservation and sustainable use of forest resources, must be viewed in the context of current global realities, country circumstances, and the overall development goals and aspirations of developing countries. The Bank's strategy needs to be cognizant of, and responsive to, the inevitable forces of globalization and their impact on forests and forest-dependent peoples. It needs to be flexible enough to accommodate local circumstances. It must also recognize the importance of the participation of constituencies within specific borrowing countries in discussions of activities such as the demarcation of forest types by function, or tree planting to supplement natural forest resources and services, and of such issues as social justice.

Revising the Bank's forest strategy and policy—and supporting implementation strategies—should enable the Bank to play two synergistic roles.

- In its global role, the Bank would capitalize on its convening powers to facilitate partnerships that mobilize *additional* financial resources (over and above improved coordination of existing country-specific aid flows) for use in client countries, including new financing mechanisms on a scale large enough to achieve any global goals that may be set out in the revised strategy.
- In its country-level role, the Bank would address the diverse realities in client countries, using all the instruments at its command and stressing long-term involvement, partnerships with a broad range of constituencies, learning by doing, and the exchange of experience across countries. This would require a long-term commitment, with enough resources for ESW and with consultative processes complementary to, but independent of, lending operations.

Recommendations

OED has identified seven elements that would make the Bank forest strategy more relevant to current circumstances and strengthen the Bank's ability to achieve its strategic objectives in the forest sector.

Mobilize financing for global forest services. The 1991 Forest Strategy acknowledged the divergence between the global and national (including local) costs and benefits of conservation. It assumed, however, that Bank clients would be willing to borrow funds to meet those objectives. But in the 80 countries with declining per capita incomes, competition is intense for Bank resources for activities with more immediate benefits. It was unrealistic to assume that global environmental objectives could be achieved through the Bank's lending function alone, even though environmental consciousness in the Bank's borrowing countries has increased and conservation would produce some long-term national and local benefits. Many of the Bank's largest borrowers, including the most forest-poor and environmentally conscious countries, have indicated that they have higher priorities for the use of Bank resources than conservation investments. The 1991 strategy produced no momentum toward designing a global collaborative effort, nor did it provide mechanisms for mobilizing adequate financial resources for such an effort. The GEF, while important, is too small in scale, providing limited, term grants for biodiversity conservation. It has no mandate to offer payments for the maintenance of forest cover and related environmental services.

Recommendation: *The Bank should use its global reach to address mechanisms for and mobilization of concessional international resources outside its lending activities. These resources should be substantial enough, and on attractive enough terms, to interest developing countries. Support for Bank leadership in developing carbon and other markets (certification, ecotourism, water) is not universal, and international willingness to pay for these services is questionable. Given the Bank's increasing decentralization, the Bank will need to revisit the matrix management arrangement governing forest operations if it is to play a global role in advancing or implementing international agreements or piloting new approaches.*

Forge international partnerships. Implementation experience suggests that the Bank cannot achieve results in the forest sector alone, even at the national level. It is nearly impossible to do so globally without forging partnerships with donors, foundations, the private sector, civil society, and NGOs. The initiatives introduced by the World Bank since 1991 represent a clear break from the past, but the Bank needs to widen the scope of its activities—not just at the international level to successfully scale up, but also at the national and local levels to scale down.

Recommendation: *The Bank needs to proactively establish partnerships with all relevant stakeholders to achieve its country- and global-level goals. The Bank and other development partners must increasingly work together at both international and country levels, in a participatory manner, to improve forest management in all kinds of forests, aiming to reach a balance among environmental, economic, and social objectives. This may lead to new agreements and new ways of mobilizing resources, with the Bank as one of many partners.*

Broaden the types of forest covered. The boreal and temperate forests of the Europe and Central Asia Region are a major source of timber, biodiversity, and other forest products and services, including recreation and carbon sequestration. Since 1991, this Region has seen the greatest growth in Bank forest lending, as countries from the Region with extensive forests have joined the Bank. Committed to protecting old growth forests, the Bank is already supporting production and conservation activities in many of these national forests, which have a strong tradition of responsible forest management for multiple uses. Such efforts to improve production efficiency in all types of forests should continue, except in forests designated for protection by national governments. It is also important to note that extending the current ban on Bank financing for commercial logging in tropical moist forests to the forests of Europe and Central Asia would jeopardize current operations and could have a chilling effect on the Bank Group's ability to mobilize much-needed funding for continued responsible forest management in this Region. Tropical dry forests are also important, especially for meeting the fuelwood and livelihood needs of the poor, particularly in Africa, which needs increased Bank support. Global thinking about the functions of forests has reached the point where it may now be possible to assign specific functions to individual forests in a way that both national and international goals can be met.

Recommendation: *Bank strategy should have a more eclectic and inclusive approach with a global reach, rather than narrowly focusing on tropical moist forests. Tailoring forest strategy to specific forest types, functions, and services would increase the Bank's global impact in the forest sector. A revised strategy should address the challenges of endangered, biodiversity-rich forests in a variety of ecosystems and should promote tree planting.*

Foster sustainable development objectives. The powerful forces of globalization and economic liberalization have intensified pressures for forest production and land conversion, challenging the goal of "sustainable development." Rapidly growing domestic and international demand for forest and agricultural products has a synergistic relationship with poor governance. Managed-production forests, tree planting, and tree plantations can reduce pressure on natural forests set aside for preservation. Together they offer the potential for "win-win" outcomes, with better yields and more conservation. The debate on prudent forest management to maintain the resource's potential into the future has advanced considerably since 1991. The improved, or low-impact, management of natural forests is recognized as holding the potential to increase the efficiency of forests' multiple functions and services. Certification is an important instrument for encouraging better practices. A number of experiments are currently under way on certification (including some associated with the World Bank/WWF Alliance), but developing countries have little experience with this instrument. Moreover, although it may be possible to agree on uniform international criteria for all processes and types of forest, uniform indicators are unlikely because of the diversity of forest types, values, and functions. So it would be unwise for the Bank to endorse a specific certification standard. The Bank must actively entertain alternative methods of certification, provided they meet generally accepted criteria and indicators and are adapted to circumstances in specific developing countries. Little is known about the economic and financial returns to improved or low-impact forest management. What evidence there is suggests that current returns to conventional logging—legal or illegal—are so high that investments in improved management cannot be justified without substantial additional research and experimentation to demonstrate its feasibility under highly diverse tropical forest conditions. In any case, widespread illegal extraction makes it pointless for entrepreneurs to invest in improved logging or tree planting. This is a classic case of concurrent government and market failure.

Recommendation: *The Bank Group should ensure that forest concerns receive due consideration in its macroeconomic work and all relevant sectors and should support activities that will help protect natural forests of national and global value. The Bank should explicitly cover approaches to these issues in its CASs and structural adjustment lending. It should increase support for quality ESW by providing resources (independent of lending operations), it should support research and extension, and it should establish guidelines, criteria, and indicators for improved forest management. Through partnerships, the Bank should also help create public and private capacity for widespread application of improved forest management and tree planting (through small, medium-size, and large community, private, and public plantation forests as appropriate in particular circumstances and with due environmental and social impact assessments in place).*

Curtail illegal logging through improved governance. The pervasiveness of illegal logging is a joint outcome of high economic returns and poor enforcement of laws and regulations. Poor governance, corruption, weak enforcement capacity, and political alliances between the ruling elite and some parts of the private sector all play a part in deforestation by permitting illegal logging and the environmentally destructive and socioeconomically inequitable exploitation of natural capital. Successful conservation, preservation of biodiversity, and improved forest management all require reducing and controlling illegal logging. Increasing forest productivity and providing alternative sources of timber and other forest products will also reduce the returns to illegal logging. Institutions that improve governance, productivity growth, and tree production are classic and global public goods and merit public investments. Each requires long gestation periods and entails the risk that results might not be achieved or might be reversed. Fortunately, the movement toward democratization and the emergence of active civil societies are paving the way for greater transparency and accountability.

Recommendation: *A revised and strengthened forest strategy should aim to reduce illegal logging by actively promoting improved governance and enforcement of laws and regulations. This would entail helping governments improve implementation of existing laws and regulations and, where necessary, changing them, improving government enforcement capacity, and diversifying sources of monitoring to actively include civil society. Improved governance cannot be achieved by the forest sector alone, but the forest sector can take the lead in bringing it about.*

Apply a more inclusive definition of "the forest-dependent poor." Depending on how "dependence" is defined, as much as a quarter of the world's poor may depend on forests for their livelihoods. Many, but not all, of the forest-dependent poor are indigenous people, a group to whom the Bank's forest strategy and related safeguards paid special attention. The 1991 strategy stressed the importance of reducing poverty to relieve pressure on forests, and promoting tree planting as a way to meet the fuelwood needs of the poor, but did not recognize the importance of developing the forest sector as a means of alleviating poverty among *all* forest-dependent people. Bank-financed projects in China, India, and Mexico demonstrate the substantial potential that forest development—through community participation—holds for generating employment, incomes, and social capital. Forest development and forest policy should become more prominent elements in the Bank's poverty alleviation strategy.

Recommendation: *Given the Bank Group's poverty alleviation mission, a revised forest strategy should include elements that directly address the livelihood and employment needs of all poor people, while continuing to safeguard the rights of indigenous people. A revised strategy should also acknowledge the frequent conflicts between the interests of the indigenous poor and those of the non-indigenous poor. To better address the needs of the forest-dependent poor, the Bank should encourage grassroots investigations into complex land rights and other rights. It also needs to monitor the impacts of macroeconomic and other changes, and develop safety nets for those likely to be harmed by Bank-supported activities.*

Adjust internal Bank incentives and reporting systems. The forest strategy's emphasis on "doing no harm" increased public accountability, made the challenges of involvement in the forest sector more complex, raised transaction costs (without raising the resources to deal with them), and was seen as increasing the reputational risks for Bank Group involvement in the forest sector. All these costs, plus the costs entailed in a more country-driven orientation, were higher than the 1991 Forest

Strategy envisaged. Against this reality, the framework of internal Bank incentives is currently tilted against forest operations, and the Bank's capacity in the forest sector has declined. Some skills, such as for assessing the impact of global, macroeconomic, and technological changes on forests, were always in short supply and remain so. These and other factors discussed in this report have made Bank managers risk-averse. Country managers also tend not to be motivated to incur the risks and transaction costs associated with complex, controversial forest operations. The Bank has not provided the resources needed to track the progress of forest operations (locally or globally) and is weak on monitoring compliance with safeguard policies.

Recommendation: *To be credible, the Bank must either align its resources with its objectives in the forest sector or scale down its objectives. The Bank's internal incentives and skill mix for forest sector operations need to be enhanced through a more evenly balanced matrix management structure so that operational staff feel that they have management's support and confidence. Staff should also have access to the necessary quality human and financial resources—if necessary, independent of lending operations—to address risky and controversial issues in the forest sectors of their client countries. If the revised strategy includes specific international forest goals, it should provide for the specific financing mechanisms and arrangements needed to achieve those goals. The Bank must diligently and routinely monitor compliance with all safeguard policies in its investment and adjustment lending, adhere strictly to the requirement for environmental impact assessments in its sectoral adjustment operations (introduced in May 1999), and consider introducing a requirement for environmental impact assessments in adjustment operations where such impacts are likely to be significant.*

ANNEXES

ANNEX A: EXPANDED EXECUTIVE SUMMARY

In 1991 the World Bank published *The Forest Sector: A World Bank Policy Paper*, a comprehensive statement of the World Bank Group's forest strategy.[1] This paper brought the environmental agenda into the mainstream of the Bank's activities and challenged the Bank Group to adopt a multisectoral approach that would conserve tropical moist forests and expand forest cover. It also initiated participatory and consultative processes in strategy formulation. Parts of the strategy outlined in the 1991 paper became the basis for Operational Policy 4.36 and Good Practices 4.36, both issued in 1993. Two previous evaluations of the forest strategy—the 1994 *Review of the Implementation of the Forest Sector Policy* by the (then) Agriculture Department and the 1996 *Forestry Portfolio Review* by the Agriculture Department and the Quality Assurance Group—concluded that it was too early to assess the impact of the forest strategy. So this report is the first comprehensive evaluation of the Bank's forest strategy.

The Bank has implemented its 1991 Forest Strategy only partially, and mainly through an increased number of forest-related components, especially in environment sector lending. Although the strategy sent a strong signal about changed objectives in the forest sector and included a new focus on conservation, its implementation has fallen short.

Direct lending for the forest sector was not well-enough incorporated into the Bank's rural development or poverty alleviation strategies, even though many poor people and minorities in borrowing countries rely greatly on forest products and services for their livelihoods. The strategy and safeguard policies helped shift Bank-financed investments away from the kinds of projects that had previously contributed to deforestation, but they also inhibited risk-taking. Furthermore, the multisectoral approach and international cooperation that the forest strategy emphasized were not actively pursued. There was only limited integration of the forest strategy into Country Assistance Strategies (CASs), macroeconomic and sectoral analyses, and adjustment, infrastructure, and agriculture lending. In addition, rapid globalization, technological changes, and governance issues in the forest sector, as well as changes in the Bank itself over the past decade, have rendered the 1991 strategy only partially relevant. At the same time, initiatives introduced by World Bank President James Wolfensohn, such as the CEO Forum and the World Bank/World Wide Fund for Nature (WWF) Alliance, went beyond the prescriptions of the 1991 strategy.

The effectiveness of the 1991 strategy has been modest, and the sustainability of its impact is uncertain. The strategy had several inherent limitations.

First, it focused narrowly on 20 tropical moist forest countries and neglected other biodiversity-rich types of forest that are even more endangered, more important globally, or more in need of conservation. Second, while it diagnosed the problem of externalities, it did not encourage development of a mechanism for mobilizing grant or concessional funding to compensate those who conserve forests of global value, implicitly assuming that governments would borrow Bank funds to achieve global conservation objectives. Third, the strategy failed to address governance issues beyond economic solutions such as the length and price of concessions as incentives for conservation. Fourth, its consultative process was not broad enough to elicit ownership among key stakeholders and government officials in borrowing countries. Moreover, the Bank had no implementation strategy and devoted too few resources to deal with the high transaction costs of Bank involvement in the sector. Fifth, the strategy insufficiently diagnosed the powerful impacts of globalization and economic liberalization—factors external to the forest sector—on rates of deforestation and on forest-dependent people. These limitations, combined with the Bank's cautious approach, had a chilling effect on Bank involvement in improving forest management, particularly in forest-rich countries, and even in those regions of forest-poor countries that use their forests for economic development.

A revised Bank forest strategy and policy—and a Bank implementation strategy—should enable the Bank Group to play two synergistic roles:

- In its global role, the Bank would capitalize on its convening powers to facilitate partnerships that mobilize *additional* financial resources (over and above improved coordination of existing country-specific aid flows) for use in client countries, including new financing mechanisms on a scale large enough to achieve the global goals set out in the revised strategy.

- In its country-level role, the Bank would recognize and address the diverse realities in client countries using all the instruments at its command and stressing long-term involvement, partnerships with a range of constituencies, learning by doing, and the exchange of experiences across countries. This would require a long-term commitment by the Bank, with enough resources for research, economic and sector work, and consultative processes complementary to, but independent of, its lending operations.

OED has identified seven elements that would make the Bank forest strategy more relevant to current circumstances and strengthen the Bank's ability to achieve its strategic objectives in the forest sector:

1. *The Bank needs to use its global reach to address mechanisms for and mobilization of concessional international resources outside its lending activities.* Measures such as the Prototype Carbon Fund and other concessional financing mechanisms should be pursued to compensate countries that are producing forest-based international public goods such as biodiversity preservation and carbon sequestration.
2. *The Bank needs to be proactive in establishing partnerships with all relevant stakeholders to fulfill both its country and global roles.* At the same time, it must recognize the implications, in terms of additional resources, of meeting global objectives and of using participatory approaches.
3. *The focus on primary tropical moist forests needs to be broadened to encompass all types of natural forests,* including temperate and boreal forests and other highly endangered, biologically-rich types of forest: the *Cerrados* region and Atlantic forest of Brazil, the tropical dry forests of Africa, and the Western Ghats of India. The revised strategy should recognize that natural forests alone need not serve all forest functions. Some important functions (such as meeting export and urban demand, providing environmental services, and meeting the employment and livelihood needs of the poor) can as easily be served by tree planting, the expansion of which could also relieve pressure on natural forests.
4. *Forest issues need to receive due consideration in all of the Bank's relevant sector activities and macroeconomic work, and the Bank should support activities that will help protect natural forests of national and global value.* Efforts to promote forest conservation and development should be streamlined and aligned with the overall development goals and aspirations of the Bank's client countries. The synergy between development and conservation objectives needs to be recognized and actively promoted through tree planting on degraded forest and nonforestlands, energy substitution, end-user efficiency, research, technology, and dissemination.
5. *Illegal logging needs to be reduced by actively promoting improved governance and enforcement of laws and regulations.* This will require helping Bank borrowers improve, implement, and enforce existing laws and regulations. It will also require that national stakeholders (especially civil society and the private sector) demand, implement, and monitor improved governance practices.
6. *The livelihood and employment needs of* all *poor people should be addressed, while continuing to safeguard the rights of indigenous people.* More attention needs to be paid to how the forest strategy affects *all* poor people, and particular attention should be paid to the conflicting needs of different user groups.
7. *The Bank needs to align its resources with its objectives in the forest sector.* The Bank's internal incentives and skill mix need to be enhanced so that operational staff feel they have the support and confidence of Bank management and country borrowers and access to the human and financial resources needed to address the risky and controversial issues of the forest sector. The Bank must also diligently and routinely monitor compliance with all safeguard policies in its investment and adjustment lending.

Key Findings

Patterns of forest sector lending

At $3.51 billion in the 1992–99 period, overall forest sector lending was 78 percent higher in nominal terms than in 1984–91. As before 1991, it was just below 2 percent of total Bank lending for the period.

Direct lending for the forest sector has stagnated. The growth in total forest lending is explained almost exclusively by the increase in forest components of non-forest-sector projects. The integration of forest sector lending into the environment and natural resource management sector has been a positive development. These projects are largely responsive to the intent of the 1991 Forest Strategy and support protected areas and biodiversity conservation, zoning, indigenous people's rights, and community participation. But investments that directly address mainstream issues in the forest sector (involving long-term policy and institutional changes needed to improve the management of forest production or regeneration on public lands) have been hobbled by low borrower demand, high transaction costs, and the fear of public controversy through "guilt by association" with poor forest practices of public forests in borrowing countries.

Mixed results from Bank policies
The Bank strategy's cautious approach to forest management and its ban on Bank financing of commercial logging in primary tropical moist forests has been strategically and symbolically important. The Bank has not been associated with (or criticized for) wasteful and illegal deforestation and degradation, as it was before 1991. The Bank has also been more cautious in financing major infrastructure investments that are likely to harm forests.

Environmental impact assessments have improved projects at entry. The implementation of safeguard policies, while imperfect, has increased Bank accountability to a civil society that increasingly participates in environmental and social monitoring. This process has also increased the sensitivity of developing country governments to environmental concerns. Indeed, a growing number of stakeholders in borrowing countries now look to the Bank for assistance in improving their domestic capacity to ensure implementation of their own safeguard policies. These are all important process achievements and developments.

It is essential to apply "do no harm" policies to Bank lending. But implementing these policies has been largely irrelevant to the rates of deforestation. Relative returns to conventional logging and alternative land uses have been high. The synergistic relationship of the countries' own infrastructure investments, developments in agricultural technology, and market pressures to convert land to other uses, stemming from the powerful impacts of economic liberalization, globalization, technological change, and devaluation have all been key factors. Moreover, the do-no-harm approach has made the Bank Group wary of getting involved in experiments to improve forest management, to address illegal logging, or to improve the interface with forest industries, even though borrowers are strongly urging the Bank Group to help them modernize their antiquated forest sectors. Hence, the Bank has often lost the opportunity to improve the management of forests that are already being exploited in an environmentally and socially unsustainable manner.

Containing deforestation in tropical moist forests
The 1991 strategy had the dual objectives of conserving tropical moist forests and planting trees, but Bank influence on containing rates of deforestation in tropical moist forests has been negligible in the 20 countries identified for Bank focus.

The 1991 strategy underrated the "developmental" function natural forests serve in forest-rich countries. Developing countries need to increase incomes, employment, and exports to meet their development objectives. Countries rich in forests but poor in capital and budget resources have tended to use their natural capital to create and sustain livelihoods and to finance development, thereby producing a conflict between national interests and global environmental objectives.

Devolution of power to the local level has increased pressure on forests in situations where power relationships are unequal and where the income, employment, and revenue needs of local governments and their politically powerful constituents increase that pressure. Sustainable development and devolution may be necessary or desirable in the long run, but many of the costs of forgoing the financial and economic benefits of forest exploitation are local and immediate, while the environmental benefits of forest conservation are national and global—and even where they are local, are either long-term or accrue to groups with only a limited voice in local governance. The situation can improve only if countries develop broad-based participatory institutions that offer equal voice to local constituencies, especially the vulnerable, and only if they can find the resources to forgo the short-term exploitation of forests.

Forest-rich countries have tended to exploit their resources inefficiently. In many countries illegal logging in natural forests accounts for at least half of the total timber supply, and wastage through conventional logging, processing, and transport is reported to be as high as 70 percent of the total harvest.

The Bank strategy adopted the ban on financing for commercial logging on the defensible grounds of uncertain valuations of forest environmental services, inadequate forest management systems, and irreversibilities associated with forest loss. But poor governance, corruption, and political alliances between various segments of the private sector and ruling elites, combined with minimal local and regional enforcement capacity, have played a part in environmentally damaging and socioeconomically inequitable exploitation of natural capital.

Bank strategy should address these issues through increased and improved partnerships with local governments, civil society, and the progressive private sector, rather than assuming that the absence of Bank Group financing for commercial logging in primary tropical moist forests will somehow reduce deforestation.

The poor have been less a source of deforestation and forest degradation in the forest-rich countries than the 1991 strategy assumed. A much stronger factor appears to have been the growing domestic demand for wood energy in industry, for timber in construction, and for tropical forest products for export to international markets. To contain deforestation and degradation, countries need substitutes. For example, investments in environmentally friendly alternative sources of energy, and tree planting could help reduce demand for wood energy from natural forests. Without such investments, and without adequate finances to compensate countries that incur costs to achieve global environmental objectives, deforestation and forest degradation will continue.

Financing mechanisms

The 1991 strategy recognized the need for international transfers to underwrite conservation of global value, but failed to generate the momentum needed to establish adequate mechanisms or finances for that purpose. There is little borrower demand for Bank funds to finance pure conservation, through protection of existing natural forests, primarily because short-term domestic economic or social returns to conservation are limited, even in countries where environmental consciousness and awareness of environmental services—such as ecotourism—have increased. Demand has also declined for the expansion of forest cover by tree planting, which could serve many forest functions and could help relieve pressure on natural forests. The long time needed for tree growth (when there is no cash flow and when risks are high), combined with tight credit, high domestic interest rates, and the continued supply of timber from natural forests, has thwarted demand for investments either in small-scale tree planting or in plantation forests. Whether the cause has been resistance to plantations on environmental and social grounds or financial policy reform restricting directed credit to specific activities, tree planting has been neglected. Grant resources provided by the Global Environment Facility are too small relative to the need. Difficulties in measuring and valuation persist, but enough is now known for the global community to move ahead with tree planting and with measures to provide financial support for the conservation of forests of global and national value. Unless adequate compensation is provided for those involved in forest conservation, forests will continue to shrink until scarcities become acute enough to elicit policy and market responses. Until adequate resources are available to provide such compensation, the Bank should resist pressures to exhort governments to achieve global objectives that cannot be justified solely on the basis of country benefits and borrower demand.

Changing country policies

Bank leverage through policy conditionality is more limited than popularly believed, even though environmental awareness among borrowers has increased. A few of the new International Monetary Fund (IMF) stabilization and Bank adjustment packages have incorporated forest sector conditionalities, but at the same time, stabilization and adjustment policies have placed caps on government spending, including spending for enforcement. Proponents argue that IMF and Bank conditionality helps bring forest policy to the attention of high-level policymakers. Indeed, useful "stroke-of-the-pen" reforms have been achieved in a few countries facing financial crisis. But externally induced reforms without domestic commitment and strong domestic champions tend not be sustainable. External pressure to achieve policy reform during a period of government weakness lacks legitimacy and, over time, may even solidify resistance. What is needed is a long-term, comprehensive political-economy approach based on nurturing domestic ownership of reforms and using all the instruments at the Bank's disposal, including lending, Country Assistance Strategies (CASs), and economic and sector work (ESW). In a long-term strategy that combines analysis, constituency building, and resource monitoring, adjustment lending with specific forest conditions can have a legitimate place in specific countries—but it is not a panacea.

The treatment of environmental and social impacts in the Bank's rapidly expanding adjustment lending is a more disconcerting matter. Macroeconomic adjustment is often essential to sustain economic growth and alleviate poverty, but its short- and medium-term impacts on forest cover and quality—and on the lives of forest-dependent people—can be devastating. The Bank needs to better understand the potential impacts of its adjustment lending—by encouraging and supporting research—and to consistently apply safeguard policies to all its lending.

To develop a framework of mutual responsibility and accountability for improved forest management, the Bank needs high-quality research and ESW at the country level. During project preparation and implementation, the Bank should build borrowing countries' capacity for forest management and forge long-term links with domestic constituencies that can conduct analysis, stimulate reform, and ensure its sustainability. This "constituency-building" approach contrasts sharply with the "conditionality" approach set forth in OP 4.36 and calls for a truly multisectoral, long-term effort, as is intended in the Bank's Comprehensive Development Framework (CDF). It also requires more resources. Internally, the Bank has often failed to incorporate forest concerns into its CASs or to effectively link CAS pronouncements with its operations. Under the current conservation-oriented forest strategy, in an increasingly country-driven and decentralized Bank, borrowers in forest-rich countries have been reluctant to include forest concerns in their own priorities for Bank support (which is part of the reason the sector gets little attention in CASs). With a few recent exceptions, forest issues have not been emphasized in ESW, even in countries where forest sectors are economically important.

Improving tree cover, sequestering carbon, and protecting biodiversity in other forests

Although the 1991 forest paper focused on tropical moist forests, Bank operations have rightly outpaced that strategy. The greatest growth in forest lending has been in the management of public forests in Eastern and Central Europe, the Bank Region with the world's largest forest cover. The boreal and temperate forests in that Region represent a major source of timber, biodiversity, forest products, and forest services, including recreation and carbon sequestration. Committed to protecting old growth forests and uninhibited by the 1991 Forest Strategy's restriction on financing commercial logging in primary tropical moist forests, the Bank is supporting production and conservation activities in many of these multiple-use forests. Such efforts to improve the production efficiency of all types of forests should continue, except in forests national governments designate as protected. Boreal and temperate forests and other forest types in the Bank's borrowing countries have their own management, conservation, and biodiversity issues that need to be addressed. But the treatment of these issues has been eclipsed by the 1991 strategy's focus on tropical moist forests. The Bank and the Global Environment Facility support them only through small operations.

The Bank's forest strategy must proactively learn from the experience of its borrowers, rather than remaining top-down and externally driven. Some forest-poor developing countries are ahead of the Bank in their forest policies and innovative approaches. Some of them are developing national sources of financing for environmental actions of national interest. The spread of democratization and increasing demands for transparency and accountability, as well as vibrant nongovernmental organization (NGO) movements in the Bank's borrowing countries, have collectively increased opportunities for impact. The Bank's convening power, its policy advice, its lending, and the prestige of its association are important in countries beginning to commit to improving forest sector management. Forest cover is stabilizing in China and India, for example, where the Bank is helping to operationalize participatory approaches to forest management. In Costa Rica, a multisectoral approach to policy advice facilitated the implementation of far-reaching reforms and national financing mechanisms. In Brazil, Cameroon, and Indonesia, where the Bank has had difficult relations in the forest sector, dialogue has now opened up both within the countries and between the countries and the donor community. This opening-up has increased the likelihood that reforms generated by constituencies within these countries will have broad ownership. In contrast despite a long-term Bank presence, in Kenya, many domestic and donor accomplishments have been lost for lack of government commitment.

Need for a poverty focus

Most of the population that lives in and around forests is among the poorest and often includes indigenous minorities. Estimates vary widely, depending on the concept of forest dependence used, but this population may include as much as a quarter of all poor. Nearly four-fifths of the forest-dependent poor are in Africa, East Asia, and South Asia. How the loss of forest cover, forest degradation, and tree planting affect these people depends on the nature of their forest dependence. Knowledge about the nature of their dependence is crucial to the judicious design of interventions. Some of the poorest forest-dependent people are women, for example, but gender considerations have not received much attention in the implementation of the Bank's strategy. Given the complex cultural challenges of getting women involved in poverty-oriented forest projects, the Bank should make extra efforts to consult with and engage nationals of borrowing countries who are knowledgeable in these issues. Bank-financed projects should involve the entire spectrum of forest-dependent people.

While many of the forest-related poor are indigenous, even larger groups of non-indigenous people have benefited from government land policies that have sometimes arbitrarily withdrawn land rights from one group and assigned them to others. The Bank's 1991 strategy underestimated the complexities of local circumstances and the need for location-specific solutions suited to the political, cultural, legal, and economic context. Bank projects have often tried to tailor the Bank's universal forest and safeguard policies to specific country conditions, but frequently the net effect has been increased transaction costs and criticism from both borrowers and outside observers. Even in countries with large amounts of forest lending, forests have not been an important element of the Bank's strategy for poverty alleviation.

Fortunately, noteworthy changes are under way in sectoral analyses and adjustment lending in the East Asia and Pacific Region, and in increasingly sophisticated participatory approaches in East Asia, South Asia, Latin America, and Eastern Europe. Project design has improved through the use of phased and more participatory approaches, including learning by doing. These important new pilot approaches need to be promoted, but it is too early to assess their outcomes or sustainability. The Bank's challenge will be to integrate forest sector issues more substantively into CASs and to provide the right incentives for piloting new approaches, and then scaling up those with positive outcomes.

The success of large programs in forest-poor China and India (representing 60 percent of the Bank's direct forest sector lending) suggests that through a combination of policy and institutional reforms and investments it is possible to achieve win-win outcomes—greater poverty reduction and an improved environment. Yet such investments are rarely made in Africa, where millions of poor people are forest-dependent. The investments in China and India show that forest regeneration and tree planting through community participation can offer substantial economic benefits to millions of poor households in forest-poor countries, while increasing forest cover, sequestering carbon, and reducing pressure on natural forests. But it takes a long time to develop and nurture community organizational structures, establish new rules of the game, and attain legal, environmental, organizational, and financial sustainability. Moreover, the risks to those investments are high given the conservative attitudes of most forest departments and the extreme poverty and remoteness typical of forested regions.

Can other donors provide the much-needed investments in forest-poor countries while the Bank contributes through ESW and policy dialogue? Some environmental NGOs strongly oppose *any* Bank investment operations in the forest sector, either because they increase the borrower's indebtedness or because they benefit forest ministries and departments uncommitted to reform.[2] Good ESW is necessary and adjustment lending can play a role in specific cases, but only in exceptional cases are both sufficient to achieve impacts on forest cover and quality on the ground. The Bank's presence—including its investments—is crucial to its convening power and its ability to view forest development in a macroeconomic and multisectoral context. Besides, ESW is increasingly tied to lending activities, and few donors or governments can scale up approaches to participatory forest development to the level the Bank can. Through a hands-on approach, the Bank learns from and gains the trust and confidence of local stakeholders, enabling it to be a credible partner in local and national problem solving. Yet demand from finance ministries for International Bank for Reconstruction and Development (IBRD) or even International Development Association (IDA) financing is slackening—even in countries with successful forest sector operations—because of competing demands on Bank funds from other poverty-reducing sectors that attract less controversy and criticism and have lower transaction costs than the forest sector, such as health and primary education. Moreover, fiscally strapped developing countries with declining per capita incomes are rarely in a position to borrow resources for participatory forest management; they have other, more urgent development priorities.

Without additional grant or concessional assistance, governments are unlikely to be interested in long-term Bank involvement in the risky forest sector. And international willingness to provide grant funds appears to be weaker today than when the forest strategy was formulated in 1991, despite growing environmental awareness in developing countries. To mobilize concessional resources, the Bank needs to work effectively with donors, with U.N. agencies, with the private sector, and with NGOs. The Bank has become more active in forging partnerships in the past few years, but the record on international cooperation has not been as strong, either because donor and Bank agendas differed, or because the Bank chose to go it alone.

Toward an effective forest strategy

The Bank needs to articulate its future role in global forest partnerships and support that role with resources commensurate with the challenge. The Bank's internal incentives currently tilt against forest operations. The loss of forest staff after adoption of the 1991 Forest Strategy sent a negative signal to operational staff. What incentives do country managers have to incur the risks and transaction costs associated with complex and controversial forest operations? The Bank lacks the instruments and, increasingly, the skills to implement a forward-looking forest sector strategy. Resources to track the progress of forest operations (globally or locally) are lacking, and arrangements for monitoring safeguard policy compliance are weak. Should budgetary resources for the forest sector be made available—even with low borrower demand—to meet global objectives? Without a clear operational policy consistent with the intentions of the revised strategy—and an implementation strategy to meet the policy goals—the Bank would be unwise to promise global results that exceed the sum of its country operations.

ANNEX B: METHODOLOGY OF THE REVIEW

The OED Review of the World Bank's 1991 Forest Strategy and its Implementation consists of two components: First, a review of all lending and non-lending activities of the World Bank Group (IBRD, IDA, IFC, and MIGA) and the Global Environment Facility (GEF) that are pertinent to the implementation of the forest strategy; second, six in-depth country case studies (Brazil, Cameroon, China, Costa Rica, India, and Indonesia). The relationships of the study's various parts are shown in figure B.1. The Portfolio Review of Bank and GEF activity was done by OED. The IFC review was done by the Operations Evaluation Group (OEG) of the IFC, and the MIGA review was done by the Office of Guarantees of MIGA.

FIGURE B.1. RELATIONSHIP OF THE STUDY'S PARTS

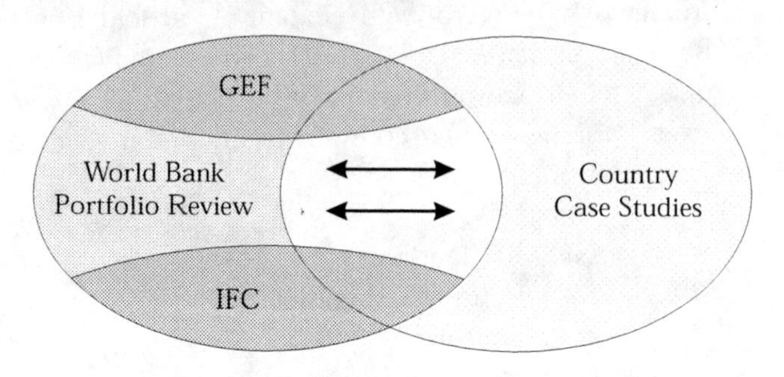

Sequence of the Review

All Bank activities in the forest sector were addressed in a multisectoral context. For analytical purposes the review concentrated on the following groups of countries:

- Six case study countries
- Twenty tropical moist forest countries
- Countries with Bank forest projects after 1991
- Countries with forest-component projects after 1991
- Countries with GEF forest projects
- Countries with forest projects before 1991
- Countries with forest-component projects before 1991.

The Bank's lending activities in the forest sector were examined in two eight-year periods, the period before implementation of the forest strategy (1984–91) and the period after implementation began (1992–99). This was done to discern whether there has been any significant change in the pattern of Bank lending since the forest strategy became effective.

Investment Lending Operations

The examination of the Bank's forest sector lending activities had three major components:

- *Direct Forest Projects:* Projects listed in the Bank's databases under the "forestry" subsector whose parent sector is agriculture.
- *Forest-Component Projects:* Projects listed in subsectors other than "forestry" that also conduct forest-sector-related activities. Forest-component projects are generally listed in other subsectors of agriculture and environment, including natural resource management.
- *Forest-Impact Projects:* All lending operations in various sectors and subsectors that have potential indirect (positive or negative) impacts on forests.

Projects with Impacts on Forests

The major sectors with forest-impact projects are agriculture, including agriculture sector adjustment lending; transportation; mining; oil and gas; and electric power and energy. Application of safeguards in these sectors was examined both for their content relative to the intentions of the 1991 Forest Strategy and their application during project preparation, approval, *and implementation.* The specific safeguards examined for compliance were OP 4.36 (Forestry), OP 4.04 (Natural Habitats), and OP 4.01 (Environmental Assessment). All impact projects in the case study countries were identified and analyzed for the application of forest-sector-related safeguards. Because of the sheer number of such projects Bank-wide, however, the portfolio analysis considered only a sample of impact projects.

The method used to identify impact projects was twofold. First, all Bank Group projects in subsectors with a potential impact on forests were identified. In the agriculture sector, for example, all projects involving agricultural extension, agricultural adjustment, agricultural credit, agro-industries and markets, annual crops, irrigation and drainage, livestock, other agriculture, perennial crops, and agricultural research were identified. In the electric power and energy sector, projects in the distribution and transportation and the hydroelectric power subsectors were considered. Projects were also identified in the oil and gas sector, in subsectors such as oil and gas transportation and exploration. In the transportation sector, projects were identified in such subsectors as highways, railways, other transportation, and rural roads. Finally, all projects in the mining sector were considered. A total of 389 projects were identified, with total commitments of $30.6 billion (table B.1). In the next step, 20 percent of these projects were randomly selected to ascertain their impact on forests. The distribution of these 78 projects by sector, subsector, and Region is presented in table B.2.

Adjustment Lending

In its review, OED addressed forests and adjustment in two ways:

- It reviewed a sample of 34 structural and agricultural adjustment loans and credits in an effort to evaluate whether and how forest issues were addressed. Specifically, the review ensured that it had covered macroeconomic and agriculture sector adjustment loans and credits in the post-1991 period, in (1) the six case study countries, (2) the 20 countries the 1991 Forest Strategy identified as containing threatened moist tropical forests, and (3) the 19 countries to which the Bank has made direct forestry project loans and therefore has knowledge of their forest sectors.[1]
- It reviewed in depth the three countries (Cameroon, Indonesia, and Papua New Guinea) for which the IMF and Bank adjustment loans included specific forest-related conditionality.

In reviewing the adjustment lending, OED asked four questions:

- Were forests mentioned at all in the loan/credit document?
- Was there a discussion of the links between adjustment reforms and forests?
- Was there a forestry component in the loan/credit?
- Was there loan/credit conditionality related to forests?

Safeguard Policies

The Bank's safeguard policies were examined, particularly in investments in sectors with a potential impact on forests—such as irrigation, power, transport—and in sectoral and macroeconomic lending. Safeguards were examined for two categories of Bank operations: (1) projects whose activities could have an impact on forests and (2) projects or project components that directly target forest management or tree planting. Projects in the first category were examined to assess their compliance with safeguards addressing forest and environmental issues (OP 4.36, Forestry; OP 4.04, Natural Habitats; and OP 4.01, Environmental Assessment). Projects in the second category, because they had explicit forest-related goals, were examined for application of the four safeguards concerning forests and local dwellers (OP 4.36, Forestry; OP 4.04, Natural Habitats; OP 4.01, Environmental Assessment; and OD 4.30, Indigenous Peoples).

Participation Issues

The OED team developed a two-category evaluative framework to determine the extent of participation in Bank projects after 1991. Projects were rated for *level* of participation and *breadth* of participation. A five-scale ranking was used to determine the *level* of participation:

0 = No participation
1 = Information sharing (one-way communication)
2 = Consultation (two-way communication)
3 = Collaboration (shared control over decisions and resources)
4 = Empowerment (transfer of control over decisions and resources).

For *breadth* of participation, projects were rated using a 5-scale ranking system:

1 = Extremely limited participation
2 = Limited participation
3 = Moderate participation
4 = High participation.

Each forest project during the two periods (1984–91 and 1992–99) was examined for the following attributes:

- Participation in project stages: Inclusion in stated project objectives; use during project design; and (where it is possible to examine) use during implementation and evaluation.

- Indicators: Inclusion of participatory indicators in monitoring and in supervision missions.
- Methods: Social assessments; beneficiary assessments; needs assessments; surveys; participatory rural appraisals; advisory groups; informal interviews; focus groups; workshops.
- Stakeholders: Communities/local rural resource users; community-based organizations/cooperatives/local institutions/associations; local NGOs; international NGOs/research institutions; indigenous peoples; women's groups; local/district/state/government representatives; commercial private sector.

Monitoring and Evaluation

The quality of monitoring and evaluation (M&E) plans at project entry was evaluated for all forest projects and for a sample of forest-component projects approved between 1992 and 1999. The review of M&E during implementation was based on a desk study of a random sample of eight projects (four forest and four forest-component projects). Projects were chosen using a stratified random sample to cover each of the Bank's Regions and to ensure the inclusion of projects that had received performance ratings of "unsatisfactory" in the Project Status Reports (PSRs). PSRs were included to better understand how effective M&E was at identifying project problems.

The M&E review focused on the following:

- Clearly stated objectives, specific to project interventions, with well-defined and appropriate indicators
- Clearly stated mid-level outputs and/or indicators of outputs, and lower-level progress indicators
- M&E arrangements, including specific office responsible for various activities
- Information, frequency of project M&E reports, and use of the information
- Institutional strengthening focused on establishing information systems or databases or expanding research findings for the forest sector.

The rating done for this review was based on key promises in the appraisal documents. The review looked for clear statements of what the project expected to achieve by the time it was completed and for unambiguous and appropriate indicators for those achievements. Statements about outputs and/or their indicators were expected to specify outcomes of lower-level outputs and processes that were germane to achieving the project objective. The review also considered whether the document contained clearly specified progress indicators—that is, a series of lower-level actions completed in a chain of linked events to permit key stakeholders to monitor progress. Based on these criteria, appraisal documents (Staff Appraisal Reports or Project Appraisal Documents) were rated as follows:

1 = Highly satisfactory
2 = Satisfactory
3 = Marginally satisfactory
4 = Unsatisfactory.

Country Assistance Strategies and economic and sector work were reviewed in different degrees of depth.

Non-lending Services

OED's review of the Bank's non-lending services included an analysis of multisectoral policy and dialogue, Country Assistance strategies (CASs), and economic and sector work (ESW). Table B.3 and table B.4 show the breakdown, by Region, of CASs, economic reports, and sector reports for sectors relevant to forestry (agriculture, forestry, energy and mining, environment, natural resource management, infrastructure and urban development, and transportation). A significant component of this review was to analyze these documents in terms of the way they directly or indirectly (for example, through policy reforms in their sectors) related to the treatment of the forest sector. The ESW, both scheduled and unscheduled, was also analyzed.

Country Assistance Strategies

The relevant "population" for this review consisted of 274 potential CAS reports when all forest, forest-component, and GEF projects (both before and after 1991) were considered. But the review concentrated on strategies for the case-study countries (30), for tropical moist forest countries (another 48 CASs), for countries with direct forest lending (another 61), and for countries with lending for forest-component projects (another 80) after 1991 for a total of 219 strategies in 65 countries.

It is important to note that this was a review of the CAS documents as opposed to a review of the effectiveness of assistance to the countries, which is addressed in the country case studies and the portfolio review. The evaluative framework specified the following elements:

TABLE B.1. WORLD BANK PROJECTS WITH POTENTIAL IMPACT ON FORESTS (1992–99)

Sector	Subsector	AFR		EAP		ECA		LCR		MNA		SAR		All Regions	
		No. of projects	Commit-ments $M	No. of projects	Commit-ments $M	No. of projects	Commit-ments $M	No. of projects	Commit-ments $M	No. of projects	Commit-ments $M	No. of projects	Commit-ments $M	No. of projects	Commit-ments $M
Agriculture	Ag. extension	12	293			2	23	2	79	1	25	3	70	20	490
	Ag. adjustment	22	707	1	50	7	1,072	6	999	4	92			40	2,920
	Ag. credit	3	149	2	272	8	136	2	345	1	121			16	1,023
	Agro-industry & marketing	5	58			9	380	1	15	1	65			16	518
	Annual crops	2	35			1	50							3	85
	Irrigation & drainage	8	304	4	311	9	243	9	909	11	1,033	6	530	47	3,330
	Livestock	4	81			1	12							5	93
	Other agriculture	10	107	4	232	7	337	5	150	6	518	1	11	33	1,355
	Perennial crops			1	70									1	70
	Research	7	244			2	27	3	222	1	14	2	74	15	581
Total		*73*	*1,978*	*12*	*935*	*46*	*2,279*	*28*	*2,719*	*25*	*1,868*	*12*	*686*	*196*	*10,465*
Electric power and energy	Distribution & transmission	7	200	10	920	8	844	4	451	1	54	3	530	33	2,999
	Hydro	5	177	2	101	3	419	2	312			2	415	14	1,424
Total		*12*	*377*	*12*	*1,021*	*11*	*1,263*	*6*	*763*	*1*	*54*	*5*	*945*	*47*	*4,423*
Mining	Mining & other extractives	8	102	1	35	6	1,941	5	345					20	2,423
Total		*8*	*102*	*1*	*35*	*6*	*1,941*	*5*	*345*					*20*	*2,423*
Oil and gas	Oil & gas transportation	1	30	3	380	3	211			1	60	1	121	9	801
	Oil/gas exploration	3	107	1	11	5	1,311			1	100	1	180	11	1,709
Total		*4*	*137*	*4*	*391*	*8*	*1,522*			*2*	*160*	*2*	*301*	*20*	*2,510*
Transportation	Highways	18	1,539	9	930	21	2,259	18	3,193	4	244	3	397	73	8,562
	Other transportation	2	32	2	130	2	268			2	57	1	24	9	511
	Railways	2	36			3	316							5	352
	Rural roads	2	85	2	83	3	63	7	580	1	58	4	506	19	1,374
Total		*24*	*1,692*	*13*	*1,143*	*29*	*2,906*	*25*	*3,773*	*7*	*359*	*8*	*927*	*106*	*10,799*
Grand total		**121**	**4,286**	**42**	**3,525**	**100**	**9,910**	**64**	**7,600**	**35**	**2,441**	**27**	**2,858**	**389**	**30,621**

Source: World Bank databases.

TABLE B.2. ANALYZED SAMPLE OF WORLD BANK PROJECTS WITH POTENTIAL IMPACT ON FORESTS (1992–99)

		AFR		EAP		ECA		LCR		MNA		SAR		All Regions	
Sector	Subsector	No. of projects	Commitments $M	No. of projects	Commitments $M	No. of projects	Commitments $M	No. of projects	Commitments $M	No. of projects	Commitments $M	No. of projects	Commitments $M	No. of projects	Commitments $M
Agriculture	Ag. extension	3	79											3	79
	Ag. adjustment	4	73	1	50	2	52	1	75	1	30			9	281
	Ag. credit	1	4			1	15	1	95					3	113
	Agro-industry & marketing	1	6			2	51							3	57
	Irrigation & drainage	2	27	1	58	1	13	2	90	2	295	1	285	9	768
	Livestock	1	20			1	12							2	32
	Other agriculture	2	12	1	5	2	21	1	23	1	42	1	11	8	114
	Research	1	15											1	15
Total		*15*	*235*	*3*	*113*	*9*	*163*	*5*	*283*	*4*	*367*	*2*	*296*	*38*	*1,458*
Electric power and energy	Distribution & transmission	2	97	2	236	2	81	1	65			1	250	8	729
	Hydro	1	10			1	35	1	300			1	65	4	410
Total		*3*	*107*	*2*	*236*	*3*	*116*	*2*	*365*			*2*	*315*	*12*	*1,139*
Mining	Mining & other extractives	1	13			1	800	1	40					3	852
Total		*1*	*13*			*1*	*800*	*1*	*40*					*3*	*852*
Oil and gas	Oil & gas transportation			1	105									1	105
	Oil/gas exploration	1	2			1	16			1	100			3	118
Total		*1*	*2*	*1*	*105*	*1*	*16*			*1*	*100*			*4*	*223*
Transportation	Highways	3	363	2	244	4	297	3	295	1	35			13	1,234
	Railways	1	20			1	120							2	140
	Rural roads			1	55	1	8	2	137	1	58	1	273	6	530
Total		*4*	*383*	*3*	*299*	*6*	*425*	*5*	*432*	*2*	*93*	*1*	*273*	*21*	*1,903*
Grand total		**24**	**740**	**9**	**753**	**20**	**1,519**	**13**	**1,120**	**7**	**560**	**5**	**884**	**78**	**5,576**

Source: World Bank databases.

TABLE B.3. DISTRIBUTION OF SCHEDULED ECONOMIC AND SECTOR WORK

	Country Assistance Strategy	Economic report	Agri-culture	Forestry	Energy & mining	Environ-ment	Natural resource manage-ment	Infra-structure & urban develop-ment	Transpor-tation
Total reports (#)	**368**	**1,060**	**277**	**24**	**120**	**65**	**9**	**288**	**97**
Scheduled reports		536	140	17	69	24	4	127	53
Africa	117	105	43	3	16	3		31	10
East Asia and Pacific	49	97	26	3	14	5	2	34	18
South Asia	23	69	13	3	9			17	3
Europe and Central Asia	66	107	26	4	22	8	1	19	10
Middle East and North Africa	26	50	12		2	3		12	5
Latin America and Caribbean	87	108	20	4	6	5	1	14	7
Case study countries	30	85	21	4	13	4	1	33	16
Tropical moist forest countries	69	125	37	6	16	4	1	31	10
Countries with Bank forest projects (after 1991)	83	154	40	5	14	6		41	19
Countries with Bank forest-component projects (after 1991)	169	295	71	10	32	8	1	77	32
Countries with GEF forest projects (after 1991)	139	231	55	10	32	8	1	70	28
Countries with forest and forest-component projects, including GEF (after 1991)	230	358	90	13	44	12	1	92	38
Countries with Bank forest projects (before 1991)	110	205	54	8	25	6	1	61	21
Countries with Bank forest-component projects (before 1991)	82	146	32	4	24	2		39	17
All countries with forest and forest-component projects (before 1991)	146	245	63	9	32	6	1	69	24
Total reports, without double-counting (before and after 1991)[a]	**274**	**397**	**103**	**15**	**52**	**12**	**1**	**96**	**38**

a. Includes Bank and GEF forest and forest-component project countries, case study countries, and 20 moist tropical forest countries, but each report is counted only once.

TABLE B.4. DISTRIBUTION OF UNSCHEDULED ECONOMIC AND SECTOR WORK

	Economic report	Agri-culture	Forestry	Energy & mining	Environ-ment	Natural resource manage-ment	Infra-structure & urban develop-ment	Transpor-tation
Total reports (#)	**1,060**	**277**	**24**	**120**	**65**	**9**	**288**	**97**
Unscheduled reports	524	137	7	51	41	5	161	44
Africa	209	61	2	28	18	1	60	17
East Asia and Pacific	53	10		4	6	1	20	4
South Asia	55	9		3	1		21	6
Europe and Central Asia	78	24	1	9	4	1	27	9
Middle East and North Africa	30	13		2			20	4
Latin America and Caribbean	99	20	4	5	12	2	13	4
Case study countries	49	8	1	7	1		18	4
Tropical moist forest countries	96	15		9	5		23	4
Countries with Bank forest projects (after 1991)	140	24		7	6		42	9
Countries with Bank forest-component projects (after 1991)	248	55	2	26	16		87	23
Countries with GEF forest projects (after 1991)	216	47	1	23	12		74	19
Countries with forest and forest-component projects, including GEF (after 1991)	308	69	2	28	14		106	26
Countries with Bank forest projects (before 1991)	180	47	2	17	13		53	15
All countries with Bank forest-component projects (before 1991)	140	27	2	15	4		41	12
All countries with forest and forest-component projects (before 1991)	250	59	3	21	15		69	17
Total reports, without double-counting (before and after 1991)[a]	**387**	**88**	**4**	**37**	**19**		**124**	**32**

a. Includes Bank and GEF forest and forest-component project countries, case study countries, and 20 moist tropical forest countries, but each report is counted only once.

- The valuable aspects of the object of evaluation
- The range of values that could be assigned to those aspects
- An aggregation formula
- A decision rule for determining an overall judgment.

The CAS review focused on the treatment of six issues:

(1) Environmental issues in general
(2) Specific forest sector or biodiversity issues
(3) Institutional development
(4) Stakeholder involvement
(5) Mechanisms for monitoring and evaluation
(6) The multisectoral approach.

The OED team saw the way these issues are discussed in a CAS as a reflection of the degree to which the country team viewed forest sector issues strategically. Therefore the ratings should be considered in the context of the Bank's overall approach to the forest sector in a country context and not in isolation. The ratings were established as follows. For issues No. 2, 3, and 4:

0 = the issue was not mentioned.
1 = the issue was mentioned but not elaborated upon.
2 = there was an elaborated discussion of the issue or it was considered a priority.
3 = goals or instruments were set for dealing with the issue.

For environmental issues in general, there were only two possible values: 0 if not mentioned and 1 if mentioned. Similarly, for M&E, 0 means there was no system to monitor outcomes in the sector and 1 means there was such a system. For the multisectoral approach, 0 means a multisectoral approach for the forest sector was not explicitly advocated in the document and 1 means such an approach was advocated.

BOX B.1. FAO DATA ON THE STATE OF THE WORLD'S FORESTS

According to the Food and Agriculture Organization (FAO), forests cover a quarter of the world's land area—about 3.5 billion hectares—and forested area diminished 0.3 percent a year between 1990 and 1995. The FAO estimates that nearly 97 percent of this forested area is in a natural state, but natural forests are rapidly being changed into semi-natural forests and plantations. This transition has already occurred in Europe, where 85 percent of the forests are now semi-natural. The shift is occurring rapidly in many developing countries.

The FAO states that only half of the natural forested area in the world is available for wood production; the rest is legally protected or harvesting is not economically viable (for lack of transportation links, markets, or other infrastructure or because of regulations or physical inaccessibility). Some of the land may be unavailable because of low productivity, but the FAO considers as much as 82 percent of the natural forest in South America unavailable because of physical inaccessibility. The OED country studies challenge these estimates, because there is so much illegal logging and because governments have increased their investments in infrastructure.

The FAO says that in addition to forested area, there are another 1.7 billion hectares covered with woody vegetation that contributes substantially to the fuelwood supply. The FAO estimates that 75 to 80 percent of the roundwood harvested in the East Asia and Pacific Region and as much as 91 percent of that harvested in the Africa Region is used for fuelwood, although the production and consumption of fuelwood are poorly tracked, because they occur largely in the informal sector.

FAO estimates of removals from natural forests are also very large. Official concessions or production statistics account for half or less of total (including export-import) consumption. But plantation forestry, which represents only 3 percent of the world's forested area, provides most of the wood sold to industry.

Simple summing of the ratings was adopted for overall ratings, which range from 0 to 12. The decision rule was that treatment was considered:

Unsatisfactory for any rating less than or equal to 6
Satisfactory if greater than 6 and less than or equal to 9
Highly satisfactory if higher than 9.

Overview of the Country Case Studies

The country case studies were intended to complement the desk portfolio reviews. Their purpose was to bring an in-depth understanding of developing countries' forest sectors and to provide the OED study with national perspectives from a range of stakeholders. Six countries were selected for in-depth study, three forest-rich, and three forest-poor. The vast differences in the six countries selected for the case studies, and the extent of the Bank's presence, are apparent from the resource endowments, the size and the pressure of population on the land, country policies and institutions, and the rates at which forest cover, policies, and institutional arrangements have changed.

Each country study was a collaborative effort between in-country authors, who have considerable field research experience in the case study countries. The OED team benefited from field visits and from wide-ranging discussions with government officials, Bank staff, and numerous other stakeholders in the country. Detailed draft terms of reference were developed to explore certain issues common across countries and others specific to certain countries. The country case studies tried to address some issues common to many countries as much as possible by using similar databases and methodologies. They also addressed issues specific to those countries and explored the Bank's interface with countries on those issues. The point was to explore the extent to which the Bank's 1991 Forest Strategy could provide the flexibility needed to deal with quite divergent country realities.

The country studies asked two key questions:

The OED study findings cast considerable doubt on the reliability of these estimates. First, the diversity of forest sector concepts, definitions, classifications, and measurements is bewildering. Second, most developing countries face financial and technological constraints that do not allow them to measure forest sector performance adequately. So national assessments of changes in forest cover and forest quality may not be reliable enough for sound cross-country comparisons. The FAO combines national forest, demographic, and income data to project current and future rates of deforestation, so the reliability of the FAO projections is questionable. However, the organization is improving and updating its forest cover estimates with a "Global Forest Resources Assessment 2000" (FRA 2000) initiative, currently in progress.

The OED review found major differences between authenticated national estimates and FAO estimates of forest cover and loss—both levels and trends—in countries as important as Brazil and Indonesia. To ensure analytical rigor, the country studies drew almost exclusively on national data and on the literature that assesses the validity of such data.

If the international community is serious about addressing issues of forest degradation, it must provide financial support to help budget-strapped developing countries conduct forest inventories, as suggested in the 1991 Forest Strategy. However, this will require a huge long-term commitment and developing countries are unlikely to buy in to such an inventory until the international stance on the extent, sources, and causes of changes in the forest cover—and their implications for resource planning and implementation—become less combative and more inclusive of issues the developing countries face. Without a mutually shared vision and a strategy that demonstrates greater political acumen and a fuller understanding of practical realities, including the international resource transfers required to achieve global objectives, the global community will achieve few useful results.

- What do we know about the sources and causes of deforestation in each country?
- How did the Bank Group (World Bank, IFC, and MIGA) and GEF interact with each other and—considering the full gamut of instruments available to the Bank Group—with the factors affecting processes of change in the forest cover?

Data Sources

The overall portfolio review for the Bank was a desk review, based on the Bank's information systems and staff interviews. A variety of in-house data sources were used, such as the Quality Assurance Group's "Projects at Risk" database, OED's project evaluation database, the Annual Review of Portfolio Performance (ARPP) database, the Planning and Budgeting Department's (PBD) database, Project Appraisal Documents (PADs), Staff Appraisal Reports, Project Status Reports (PSRs), Project Completion Reports (PCRs), Implementation Completion Reports (ICRs), and Project Performance Audit Reports (PPARs). For data on the state of the forests in the case study counties, OED relied almost entirely on national data and the literature. FAO data, the accuracy of which is debatable, were used only when nothing else was available (see box B.1).

Process of the Study

The highly participatory process used in the study was important for reaching conclusions and presenting different viewpoints. The review consulted widely with governments, development agencies, NGOs, and the private sector. The OED design paper was translated into Portuguese, French, Mandarin, Spanish, and Bahasa (languages of five of the six case study countries: Brazil, Cameroon, China, Costa Rica, and Indonesia). The country studies and the main report were translated before the OED and ESSD regional workshops.

Process-related components of the review are described next.

Advisory Committee

An advisory committee of four counseled OED on its review from the beginning. The committee met three times, offering comments and advice on the design paper, on the selection of countries for the case studies, on preliminary findings from the portfolio review, on the overall consultation plan for the study, and on the final report. Committee members were in regular contact with the OED team.

Workshops

Entry workshop. An entry workshop on the OED review was held on December 18, 1998, at World Bank headquarters in Washington, D.C. The purpose of the workshop was to solicit comments on the substance of the issues the OED team proposed to address and on the methodology proposed to ensure that the review's output would be as relevant as possible to the ongoing ESSD Forest Policy Review and Implementation Process. Participants in the workshop included the co-chair and executive secretary of the Intergovernmental Forum on Forests, bilateral donors, members of OED's advisory committee, authors of country case studies, and World Bank Group staff and managers.

Internal reviews. OED's preliminary report was discussed with Bank staff and management at a workshop on December 15, 1999, and at a meeting of the World Bank Executive Board Committee on Development Effectiveness (CODE) on December 23, 1999.

Country workshops. OED held four multi-stakeholder country workshops between November 1999 and April 2000. The India Country Workshop was hosted by India's Ministry of Environment and Forests with active input from the World Bank's country department (November 1, 1999). The China Country Workshop, hosted by the State Forestry Administration (SFA) and World Bank country office, was held on November 5, 1999. OED held a workshop in Brazil in collaboration with the government of Brazil, the Brazilian Corporation for Agricultural Research (EMBRAPA), and the World Bank country office on November 18–19, 1999. The Indonesia Workshop, co-hosted by the International Center for Forestry Research (CIFOR) and the U.K. Department for International Development (DFID), was held on April 25, 2000. The workshops were designed to allow governments and other country-level stakeholders (such as representatives from NGOs, the private sector, and academia) an opportunity to comment on their country's case study before the studies were offered to an international audience for feedback. The workshop participants discussed both the Bank's involvement in the forest sector and the underlying diagnosis and implementation of the country's forest strategy and its relation to the Bank's 1991 strategy.

The Cameroon case study (translated into French) and the Costa Rica study (translated into Spanish) were sent for comments to a wide range of stakeholders. Comments about these case studies have been published together with the final case study reports.

OED Forest Strategy Review Workshop. Forestry experts, environmental activists, industry representatives, and government policymakers met in Washington, D.C., on January 27 and 28, 2000, to discuss the findings of the preliminary OED review and to contribute ideas to a new strategy being prepared by the World Bank's Environmentally and Socially Sustainable Development (ESSD) Network. The workshop triggered a discussion among participants, who explored how to bridge competing perspectives and how to facilitate partnerships to support conservation and the sustainable and equitable use of forest resources. Summary proceedings of the workshop are being issued as a supplement to the main OED report. Where appropriate, OED reflected comments made or submitted at the workshop in its final report.

Web-Based Consultations and Information Sharing

OED's design paper was posted on an OED website in January 1999 to elicit comments from a wider international audience. Also posted on the web were the proceedings of the entry workshop held in December 1998, a matrix summarizing comments made at that workshop and OED's responses, and a summary of meetings with NGOs and a matrix with OED's responses.

The draft synthesis report and draft country case studies were placed on the web in January 2000. OED has also issued updates on the forestry study on the Internet every two months, to inform interested stakeholders.

NGO Consultations

OED's first briefing/consultation with NGOs was held on January 29, 1999, to discuss the study design paper. Subsequently several meetings were held with NGOs, and OED also participated in meetings organized by ESSD.

Focus Group Sessions

Bank staff and managers working in the forest sector were invited to a series of focus group sessions in the late winter/early spring of 1999. The sessions were held to gather insights on specific issues from country and sector managers, lead macro- and sector economists, forestry and forest-related task managers, and specialists involved in addressing issues related to the forest sector. The sessions were facilitated by Madelyn Blair of Peleri, Inc. Five focus group sessions were held between January and March 1999. Preliminary findings of the focus group sessions, which were attended by a small number of staff, were statistically validated through a staff survey.

Staff Survey

Questionnaires were sent to 100 World Bank Group staff associated with the Bank's Forestry Community of Practice. The focus of discussions included, but was not limited to, the development of Country Assistance Strategies as well as other sector work and lending to sectors that could have an impact on forests. Opinions expressed are reflected in the context of affiliations with particular Regions and Networks, but individual responses were kept anonymous.

CEO Forum Questionnaire

As part of its review, OED also issued a survey questionnaire to all members of the CEO Forum. The survey, which sought to gauge awareness of the Bank's forest strategy at the company level, asked if members were familiar with the 1991 Forest Strategy before the CEO Forum was formed. At the time the survey was issued (April 1999), the official Forum membership was 31. Membership in the CEO Forum was designed to include "various stakeholder groups whose activities, one way or another, are critical to the present and future management of the world's forests." These stakeholder groups include leading industrialists, heads of NGOs, and members of government ministries and international organizations. The Forum was created to promote the discussion of the options for reducing barriers to sustainable management in forests, mainly by promoting responsible investors' participation in production and management of these forests.

World Bank/WWF Alliance Questionnaire

On May 3, 1999, the WB/WWF Alliance questionnaire was sent to 47 World Bank and 46 World Wide Fund for Nature staff members. Staff members from both organizations were identified through the Alliance Country Team List, which designates the coordination of country and regional contact staff members. Members of the Alliance steering committee—all but a few of whom were also assigned country or regional coordinating responsibilities—were also sent the questionnaire and included in these totals.

ANNEX C: PORTFOLIO REVIEW DATA

TABLE C.1. WORLD BANK ECONOMIC AND SECTOR REPORTS, SCHEDULED AND UNSCHEDULED, ALL SECTORS (NUMBER)

Report type	1992-95	1996-99	Net change (%)
Country Assistance Strategy	179	189	6
Economic report	533	527	-1
Agriculture	173	104	-40
Forestry	18	6	-67
Energy & mining	69	51	-26
Environment	20	45	125
Natural resource management	3	6	100
Infrastructure & rural development	158	130	-18
Transportation	49	48	-2

Source: World Bank databases.

FIGURE C.1. NET CHANGE IN ECONOMIC AND SECTOR WORK (1992–95 TO 1996–99)

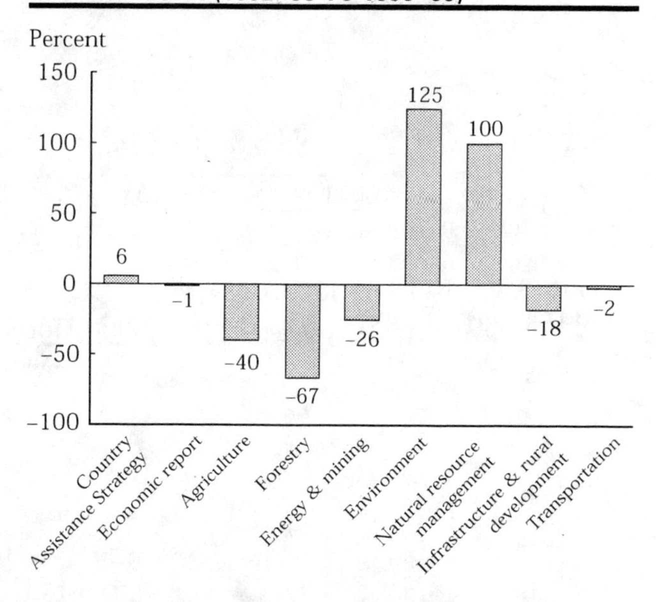

TABLE C.2. WORLD BANK LENDING OPERATIONS, BY REGION

Region	1984-91 No. of projects	1984-91 Commitments ($M)	1992-99 No. of projects	1992-99 Commitments ($M)	Change in commitments $M	Change in commitments Percent
AFR	683	23,362	589	21,685	-1,677	-7
EAP	312	30,045	363	52,317	22,272	74
ECA	132	15,060	414	33,425	18,366	122
LCR	357	38,955	472	47,570	8,616	22
MNA	171	11,073	150	10,578	-496	-4
SAR	264	30,451	187	23,760	-6,691	-22
All Regions	**1,919**	**148,946**	**2,175**	**189,336**	**40,390**	**27**

Source: World Bank databases.

FIGURE C.2. NET CHANGE IN COMMITMENTS AFTER 1991, ALL LENDING OPERATIONS

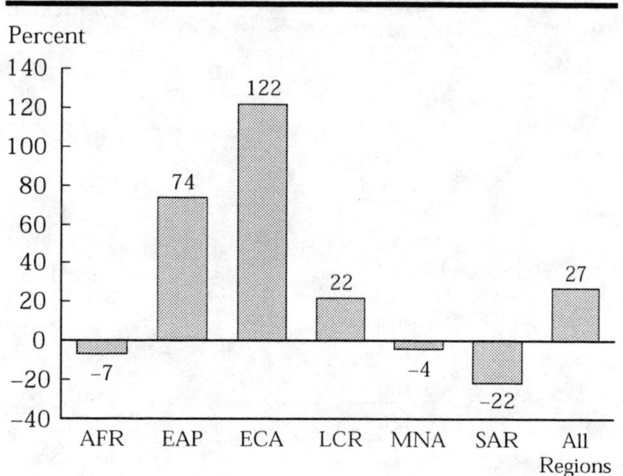

TABLE C.3. WORLD BANK ADJUSTMENT LENDING OPERATIONS, BY REGION

Region	1984-91 No. of projects	1984-91 Commitments ($M)	1992-99 No. of projects	1992-99 Commitments ($M)	Change in commitments $M	Change in commitments Percent
AFR	132	8,221	149	8,097	-124	-2
EAP	15	3,486	23	12,008	8,522	244
ECA	18	4,004	83	15,483	11,479	287
LCR	62	13,510	81	15,882	2,372	18
MNA	17	3,264	18	2,591	-673	-21
SAR	21	2,126	24	3,033	907	43
All Regions	**265**	**34,611**	**378**	**57,092**	**22,482**	**65**

Source: World Bank databases.

FIGURE C.3. NET CHANGE IN COMMITMENTS AFTER 1991, ADJUSTMENT LENDING OPERATIONS

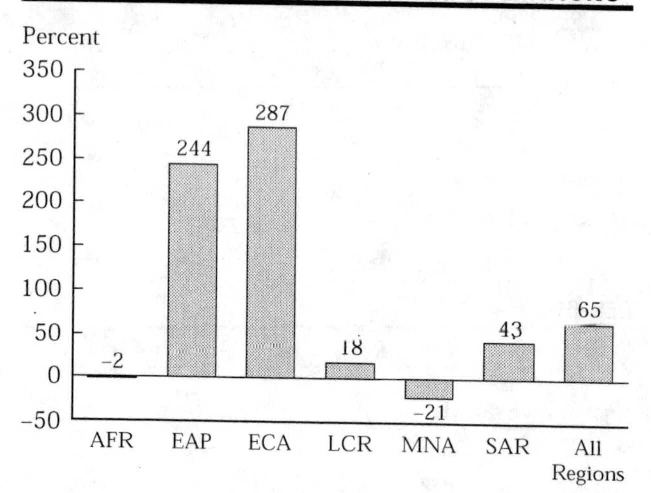

FIGURE C.4. NET CHANGE IN COMMITMENTS AFTER 1991, PRIMARY PROGRAM OBJECTIVES

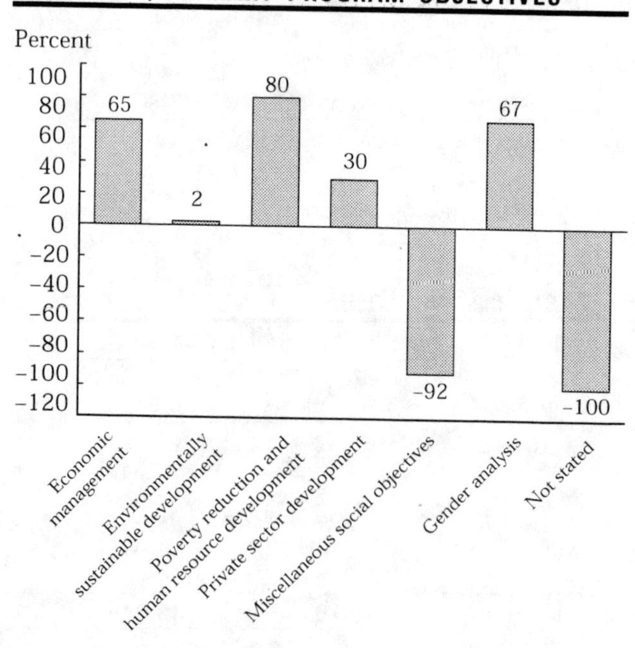

TABLE C.4. PRIMARY PROGRAM OBJECTIVES OF LENDING OPERATIONS

Primary program objective	1984-91 No. of projects	1984-91 Commitments ($M)	1992-99 No. of projects	1992-99 Commitments ($M)	Change in commitments $M	Change in commitments Percent
Economic management	399	37,879	586	62,575	24,696	65
Environmentally sustainable development	766	56,343	623	57,366	1,023	2
Poverty reduction and human resource development	397	25,280	665	45,455	20,175	80
Private sector development	199	17,907	291	23,315	5,409	30
Miscellaneous social objectives	11	1,298	1	100	-1,198	-92
Gender analysis	7	315	9	526	211	67
Not stated	140	9,925			-9,925	-100
Total	**1,919**	**148,946**	**2,175**	**189,336**	**40,390**	**27**

Source: World Bank databases.

TABLE C.5. ENVIRONMENTAL ASSESSMENT CATEGORIES, ALL BANK OPERATIONS

Environmental assessment category	1984-91 No. of projects	1984-91 Commitments ($M)	1992-99 No. of projects	1992-99 Commitments ($M)	Change in commitments $M	Change in commitments Percent
Full environmental assessment	17	2,659	165	26,819	24,161	909
Partial environmental assessment	171	15,366	695	61,136	45,770	298
No environmental assessment	893	67,584	1,075	84,118	16,534	24
Free-standing environmental project	21	1,802	18	1,124	-678	-38
To be decided	817	61,536	222	16,139	-45,397	-74
Total	**1,919**	**148,946**	**2,175**	**189,336**	**40,390**	**27**

Source: World Bank databases.

FIGURE C.5. NET CHANGE IN COMMITMENTS AFTER 1991 BY ENVIRONMENTAL ASSESSMENT CATEGORY

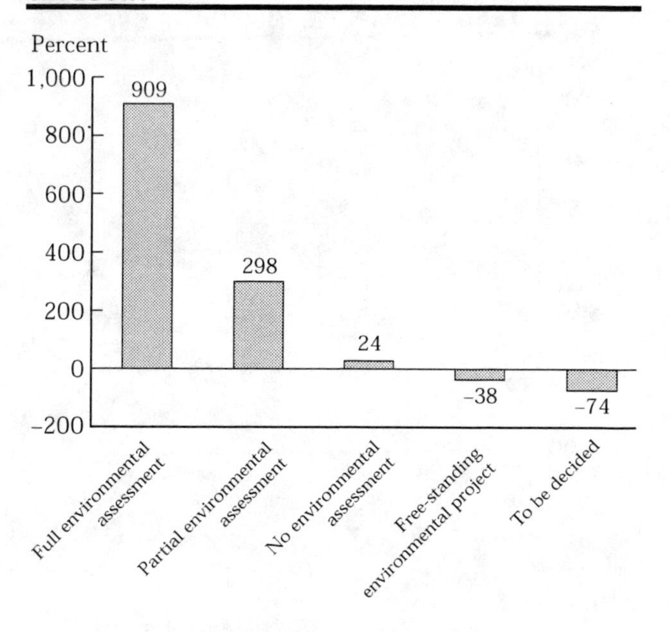

FIGURE C.6. NET CHANGE IN COMMITMENTS, BANK FOREST AND FOREST-COMPONENT PROJECTS AFTER 1991

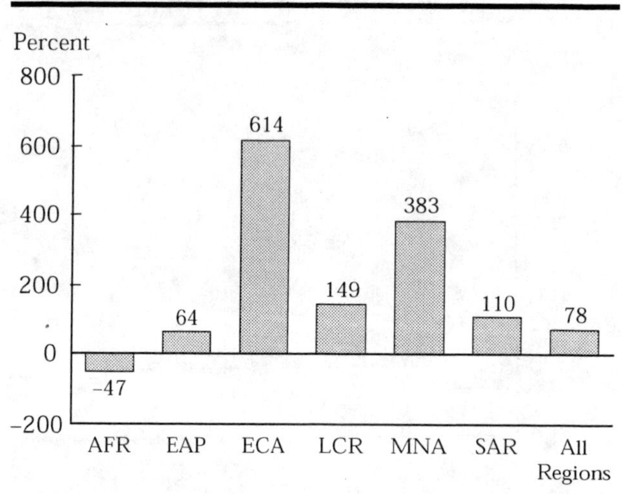

TABLE C.6. WORLD BANK DIRECT FOREST PROJECTS, BY REGION

Region	1984-91 No. of countries	1984-91 No. of projects	1984-91 Total project cost ($M)	1984-91 Commitments ($M)	1992-99 No. of countries	1992-99 No. of projects	1992-99 Commitments ($M)	1992-99 Total project cost ($M)	Change in commitments $M	Change in commitments Percent
AFR	16	17	811	426	3	3	78	53	-373	-88
EAP	5	9	1,876	723	4	6	1,059	578	-145	-20
ECA	1	1	95	35	5	5	352	245	210	600
LCR	3	3	200	103	4	5	124	86	-17	-17
MNA	2	2	150	69	3	4	1,387	221	152	220
SAR	5	9	614	327	4	11	673	540	213	65
All Regions	**32**	**41**	**3,746**	**1,682**	**23**	**34**	**3,672**	**1,722**	**40**	**2**

Source: World Bank databases.

TABLE C.7. WORLD BANK FOREST-COMPONENT PROJECTS, BY REGION

Region	1984-1991						1992-1999					
	No. of countries	No. of cost projects	Project cost ($M)	Total commitments ($M)	Forest component costs ($M)	Forest component commitments ($M)	No. of countries	No. of projects	Total project cost ($M)	Total commitments ($M)	Forest component costs ($M)	Forest component commitments ($M)
AFR	15	16	547	281	188	90	14	20	1,536	718	464	219
EAP	1	2	385	152	14	6	6	29	6,687	3,189	1,506	618
ECA	2	2	340	220	18	10	5	7	431	282	115	74
LAC	4	5	580	360	213	151	14	24	2,359	1,351	855	547
MNA							5	7	1,526	346	146	113
SAR	5	7	2,004	931	43	34	3	7	492	322	322	219
All Regions	**27**	**32**	**3,856**	**1,945**	**477**	**291**	**47**	**94**	**13,032**	**6,209**	**3,408**	**1,790**

Source: World Bank databases.

FIGURE C.7. NET CHANGE IN COMMITMENTS AFTER 1991, OPERATIONS IN 20 COUNTRIES WITH THREATENED TROPICAL MOIST FORESTS

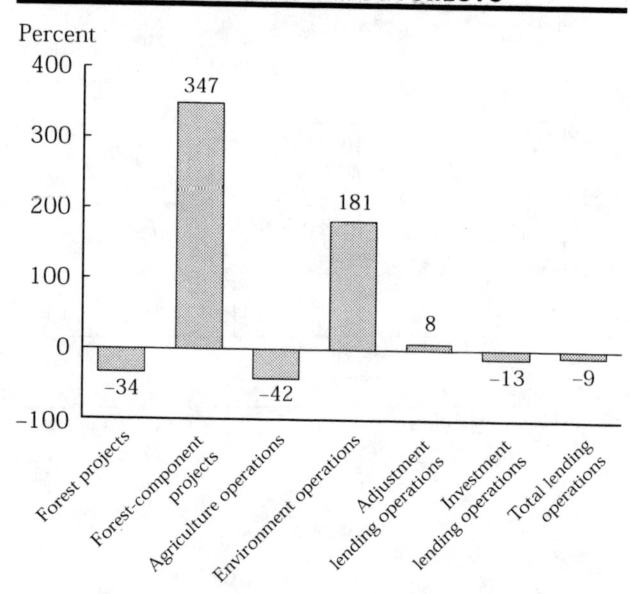

FIGURE C.8. NET CHANGE IN COMMITMENTS AFTER 1991, BANK FOREST AND FOREST-COMPONENT PROJECTS, INCLUDING GEF

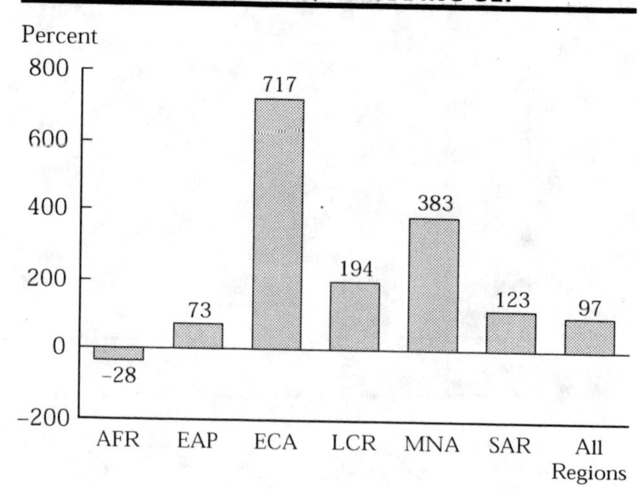

TABLE C.8. WORLD BANK FOREST AND FOREST-COMPONENT PROJECTS, BY REGION

Region	1984-91		1992-99		Change in commitments	
	No. of projects	Commitments ($M)	No. of projects	Commitments ($M)	($M)	Percent
AFR	33	516	23	272	-244	-47
EAP	11	729	35	1,196	467	64
ECA	3	45	12	319	274	614
LCR	8	254	29	633	379	149
MNA	2	69	11	333	264	383
SAR	16	361	18	759	398	110
All Regions	**73**	**1,973**	**128**	**3,512**	**1,539**	**78**

Source: World Bank databases.

TABLE C.9. DISTRIBUTION OF FOREST-COMPONENT PROJECTS, BY SECTOR

Sector	1984-91 No. of projects	1984-91 Total commitments ($M)	1984-91 Forest component commitments ($M)	1992-99 No. of projects	1992-99 Total commitments ($M)	1992-99 Forest component commitments ($M)
Agriculture	18	599	75	47	3,759	887
Electric power and energy	5	845	15	3	251	7
Environment	4	223	168	43	1,952	863
Finance	1	75	22			
Industry	1	9	1			
Public sector management	3	194	9			
Social				1	248	32
Total	**32**	**1,945**	**291**	**94**	**6,209**	**1,790**

Source: World Bank databases.

TABLE C.10. WORLD BANK LENDING OPERATIONS IN 20 COUNTRIES WITH THREATENED TROPICAL MOIST FORESTS

Project	1984-91 No. of projects	1984-91 Commitments ($M)	1992-99 No. of projects	1992-99 Commitments ($M)	Change in commitments ($M)	Change in commitments Percent
Forest projects	15	774	11	512	-262	-34
Forest-component projects	9	174	28	779	604	347
Agriculture operations	153	14,461	98	8,396	-6,065	-42
Environment operations	12	824	26	2,316	1,492	181
Adjustment lending operations	64	14,079	78	15,143	1,064	8
Investment lending operations	500	56,339	455	48,831	-7,509	-13
Total lending operations	**564**	**70,419**	**533**	**63,973**	**-6,445**	**-9**

Source: World Bank databases.

TABLE C.11. INTENTIONS OF FOREST SECTOR AND FOREST-COMPONENT PROJECTS (AFTER 1991)

Region	Projects	Humid tropical forest	Policy reform	Forest expansion/ intensification	Forest protection	Poverty alleviation	Participation	Intersectoral links	Institutional development	International cooperation	Introduction of new technologies
Forest sector											
AFR	3	1	2	1	3	2	1	1	3	2	3
EAP	6	1	4	3	3	3	4	0	5	2	6
ECA	5		5	5	3	2	2	2	5	1	5
LCR	5		5	1	3	4	5	1	5	1	5
MNA	4		4	1	2	3	4	0	4	2	2
SAR	11	2	6	7	10	11	11		11	1	11
Subtotal	34	4	26	18	24	25	27	4	33	9	32
Forest component											
AFR	20	1	12	7	10	16	18	6	19	11	18
EAP	29	9	8	24	7	24	18	1	20		16
ECA	7		3	1	4	3	4	4	7	5	6
LCR	24	11	19	6	17	18	21	6	23	11	22
MNA	7		5	5	3	3	6	4	6	1	2
SAR	7	1	4	3	3	6	5	1	7	4	6
Subtotal	94	22	51	46	44	70	72	22	82	32	70
Grand total	**128**	**26**	**77**	**64**	**68**	**95**	**99**	**26**	**115**	**41**	**102**
Percent	**100**	**20**	**60**	**50**	**53**	**74**	**77**	**20**	**90**	**32**	**80**

Note: **Humid tropical forest:** Projects that address issues related to tropical moist forests in the 20 countries identified in the 1991 Forest Strategy.

Policy reform: Includes both direct reform in laws, policies, and regulations, and activities such as ecozoning, inventories, regional diagnostic studies, and national plans for forest and biodiversity conservation that provide information for policymaking.

Forest expansion/intensification activities: Refers to activities that enhance production such as agroforestry, plantations for fuelwood, timber, crop trees (rubber, oil palm, and fruit trees), silvicultural treatments, preparation/implementation of forest management plans, and promoting natural regeneration.

Forest protection: Interventions intended to preserve intact natural forest areas or conserve biodiversity through the system of protected areas. Includes ecological zoning, biodiversity inventories, policy and regulation related to protected areas, and expansion and/or improvement of management of protected areas.

Poverty alleviation: Covers activities such as granting/enforcing tenurial rights; income-generation projects; sharing benefits of the forests with local communities; social investments such as building roads, hospitals, or schools; building skills of villagers; or supporting local organizations.

Participation: Refers to the involvement of stakeholders (national, regional, local) in appraisal and implementation. Includes information sharing, consultation, creation of advisory committees, and stakeholder collaboration in implementation.

Intersectoral links: Projects that link forest management with other sectors such as energy, tourism, and infrastructure. Includes only projects with links outside the natural resource management sector.

Institutional development: Includes institutional reforms, support to local community institutions, development of government agencies responsible for forest management through training, creation of new units, and establishing environmental information systems.

International cooperation: Includes cofinancing projects with other donors, participation in consultative groups, coordination of activities with other donors, and incorporating lessons learned from other donor projects.

New technologies: Includes introduction and adaptation of new or improved technologies to enhance production, planning, or information management.

ANNEX D: QUALITY ASSURANCE GROUP RISK RATINGS FOR ACTIVE BANK OPERATIONS

The Quality Assurance Group (QAG) maintains a database of all active World Bank projects, that identifies whether a project is at risk (that is, an actual problem project or a potential problem project) or not at risk. Project risk assessments are based on:

- The extent to which development objectives are achieved through implementation performance, as reported by Bank task managers during project supervision, typically every six months.
- The extent to which conditions agreed to during project design and negotiations are being met during implementation.
- Whether any unanticipated domestic factors (such as macroeconomic or political problems) or external shocks (such as climate or terms of trade) are affecting project performance.

The factors used to determine project performance and risk status are effectiveness delays, compliance with legal covenants, management performance, availability of counterpart funds, procurement progress, environment and resettlement, slow disbursements, a history of past problems, risky country, risky subsector, and economic management.

QAG reported that of the Bank's 1,775 active lending operations with total net commitments of $124.81 billion, as of June 1999, 77 percent of the projects were rated not at risk. Active operations include 30 forest projects with commitments of $1.68 billion and 80 active forest-component projects with total project commitments of $5.3 billion (table D.1). The percentage of forest projects not at risk is far higher, and the percentage of forest-component projects not at risk is lower than for projects in the agriculture sector, environment sector, or the entire Bank portfolio (figure D.1).

FIGURE D.1. PERCENTAGE OF ACTIVE PROJECTS NOT AT RISK, SELECTED SECTORS

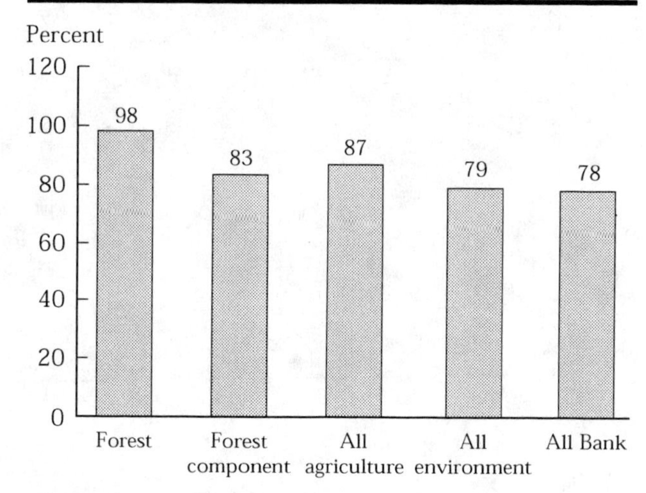

Source: QAG, June 1999.

The regional breakdown shows that only the ECA Region has any forest projects at risk, and only one of the five forest projects in the ECA Region is at risk (figure D.2).

The regional distribution for forest-component projects, in contrast, shows that only 55 percent of the projects in the LCR Region and 71 percent in the AFR Region are *not* at risk. But none of the forest-component projects in the ECA and MNA Regions is at risk (figure D.3).

The OED ratings for closed forest projects tend to be substantially lower than the QAG ratings for active forest projects. This disparity can be explained by the different methodologies used by QAG and OED. QAG combines self-assessment by project staff and potential problem ratings by supervision staff. The potential problem ratings are based on 12 "flags," 3 of which are exogenous—that is, unrelated to the characteristics of

TABLE D.1. RELATIVE RISK RATINGS FOR ACTIVE FOREST AND FOREST-COMPONENT PROJECTS (JUNE 1999)

	Active projects		Actually at risk		Potentially at risk		Not at risk	
Project type	No. of projects	Net commitments ($M)	No. of projects (%)	Net commitments (%)	No. of projects (%)	Net commitments (%)	No. of projects (%)	Net commitments (%)
Forest	30	1,683	3	2			97	98
Forest component	80	5,295	16	8	9	9	75	83
All agriculture	268	18,061	12	10	5	3	83	87
All environment	96	5,094	14	11	9	10	77	79
All World Bank	**1,577**	**124,809**	**15**	**16**	**7**	**6**	**77**	**78**

Source: QAG.

FIGURE D.2. ACTIVE FOREST PROJECTS NOT AT RISK, BY REGION

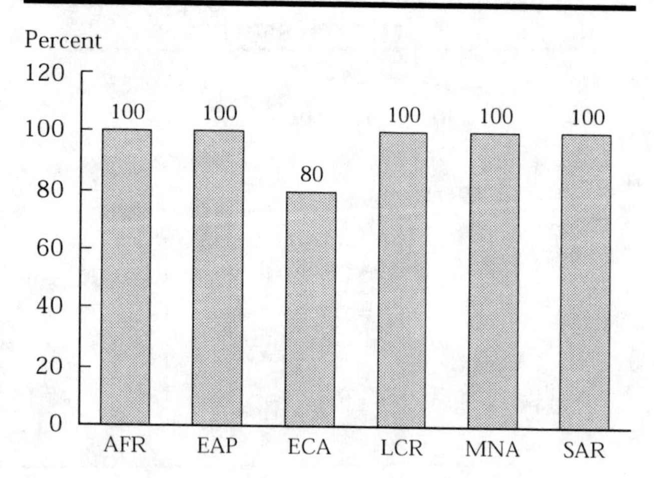

Source: QAG, June 1999.

FIGURE D.3. ACTIVE FOREST-COMPONENT PROJECTS NOT AT RISK, BY REGION

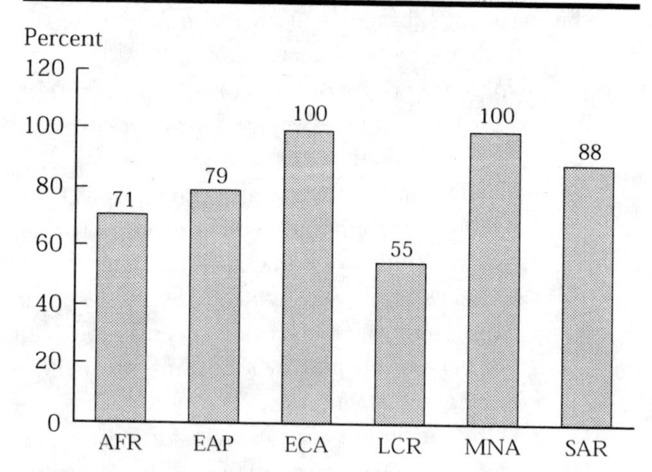

Source: QAG, June 1999.

the operation. Several other measures are drawn from sources other than the project task manager. OED, on the other hand, conducts an independent evaluation of completed projects. The ongoing forest projects may actually be performing better than past forest projects simply because they have incorporated lessons learned and OED recommendations.

"Projects at risk" is an important criterion underpinning the Portfolio Improvement Program (PIP) launched by the Bank in 1996. The PIP aims to promote substantial improvement in the Bank's portfolio performance. It complements other Bank portfolio management instruments by targeting management attention to the projects, sectors, and countries with the most severe performance problems and where intensified attention can be expected to have a high payoff. The concept of "projects at risk" was developed as a tool to target management attention to projects considered to have a high risk of failure on completion based on leading indicators known to be precursors of future problems. This work is being closely coordinated with the Annual Review of Portfolio Performance (ARPP).

Actual Problem Projects: Projects with an unsatisfactory or a highly unsatisfactory rating for Summary Implementation Progress and/or Summary Development Objectives on the latest Project Status Report.

Potential Problem Projects: Projects that are associated with at least 3 of 12 leading indicators of future problems (see table D.2).

A comparison of QAG ratings by Region for all Bank projects, as well as for all projects in the agriculture and environment sectors and for all forest projects, and forest-component projects is presented in tables D.3 through D.7. Significant risk factors associated with these sectors are presented in tables D.8 through D.12.

TABLE D.2. QAG INDICATORS OF POTENTIAL PROBLEM PROJECTS

Project flag	Description
1. Long effectiveness delays	The elapsed time between Board presentation and effectiveness is more than nine months for investment, and more than three months for adjustment and emergency operations (this flag is turned off three years after Board approval).
2. Poor compliance with legal covenants	Compliance with legal covenants rated unsatisfactory or highly unsatisfactory in the last Project Status Report.
3. Project management problems	Management performance rated unsatisfactory or highly unsatisfactory in the last Project Status Report.
4. Shortage of counterpart funds	Counterpart fund availability rated unsatisfactory or highly unsatisfactory in the last Project Status Report.
5. Procurement problems	Procurement progress rated unsatisfactory or highly unsatisfactory in the last Project Status Report.
6. Poor financial performance	Financial performance rated unsatisfactory or highly unsatisfactory in the last Project Status Report.
7. Environmental/resettlement problems	Environmental/resettlement rating of unsatisfactory or highly unsatisfactory in the last Project Status Report.
8. Significant disbursement delays	Disbursement delays of 18 months or more for investment operations or 6 months or more for adjustment and emergency operations.
9. Long history of past problems	Unsatisfactory rating on implementation progress and/or development objectives for two or more consecutive ARPPs in previous years (turned off after project restructuring).
10. In a risky country	In a country with a high "failure" rate (in which the percentage of projects closed in the previous five years and rated unsatisfactory by OED is over 50 percent, or in which the share of commitments associated with these unsatisfactory projects is more than 33 percent).
11. Poor macroeconomic setting	Located in a country with weak macro management (Development Economics Department country risk ratings—three-year average of less than 3 on a scale of 1 to 5).
12. In a risky subsector	In a subsector with a historically high "failure" rate (in which the percentage of projects closed in the previous five years and rated unsatisfactory by OED is over 50 percent, or in which the share of commitments associated with these unsatisfactory projects is more than 33 percent).

TABLE D.3. RISK RATINGS, ACTIVE AGRICULTURE SECTOR PROJECTS, BY REGION (JUNE 1999)

	Active projects		Actually at risk		Potentially at risk		Not at risk	
Region	No. of projects	Net commitments ($M)	No. of projects (%)	Commitments (%)	No. of projects (%)	Commitments (%)	No. of projects (%)	Commitments (%)
AFR	66	1,565	12	13	14	13	74	74
EAP	55	7,191	9	2	4	2	87	96
ECA	38	1,413	13	35			87	65
LCR	46	3,053	9	6	2	1	89	93
MNA	28	1,473	18	24			82	76
SAR	35	3,366	14	13	3	2	83	84
All Regions	**268**	**18,061**	**12**	**10**	**5**	**3**	**83**	**87**

Source: QAG.

TABLE D.4. RISK RATINGS FOR ACTIVE ENVIRONMENT SECTOR PROJECTS, BY REGION (JUNE 1999)

	Active projects		Actually at risk		Potentially at risk		Not at risk	
Region	No. of projects	Net commitments ($M)	No. of projects (%)	Commitments (%)	No. of projects (%)	Commitments (%)	No. of projects (%)	Commitments (%)
AFR	16	266	13	29	19	21	69	51
EAP	20	1,852	10	9			90	91
ECA	15	349	7	23	7	7	87	70
LCR	29	1,536	21	10	14	25	66	65
MNA	6	255	33	32			67	68
SAR	10	836			10	6	90	94
All Regions	**96**	**5,094**	**14**	**11**	**9**	**10**	**77**	**79**

Source: QAG.

TABLE D.5. RISK RATINGS FOR ACTIVE WORLD BANK PROJECTS, BY REGION (JUNE 1999)

	Active projects		Actually at risk		Potentially at risk		Not at risk	
Region	No. of projects	Net commitments ($M)	No. of projects (%)	Commitments (%)	No. of projects (%)	Commitments (%)	No. of projects (%)	Commitments (%)
AFR	392	14,711	19	16	15	15	66	68
EAP	287	34,636	13	11	3	2	83	88
ECA	303	21,637	17	30	8	15	76	55
LCR	322	28,333	11	10	4	4	84	86
MNA	122	6,976	16	18	3	1	81	81
SAR	151	18,516	16	15	4	4	80	80
All Regions	**1,577**	**124,809**	**15**	**16**	**7**	**6**	**77**	**78**

Source: QAG.

TABLE D.6. RISK RATINGS FOR ACTIVE FOREST-COMPONENT PROJECTS, BY REGION (JUNE 1999)

	Active projects		Actually at risk		Potentially at risk		Not at risk	
Region	No. of projects	Net commitments ($M)	No. of projects (%)	Commitments (%)	No. of projects (%)	Commitments (%)	No. of projects (%)	Commitments (%)
AFR	17	580	18	22	12	7	71	71
EAP	24	2,574	17	5	4	1	79	94
ECA	6	251					100	100
LCR	20	942	30	18	15	38	55	45
MNA	5	169					100	100
SAR	8	779			13	6	88	94
All Regions	**80**	**5,295**	**16**	**8**	**9**	**9**	**75**	**83**

Source: QAG.

TABLE D.7. RISK RATINGS FOR ACTIVE FOREST PROJECTS, BY REGION (JUNE 1999)

Region	Active projects		Actually at risk		Potentially at risk		Not at risk	
	No. of projects	Net commit-ments ($M)	No. of projects (%)	Commit-ments (%)	No. of projects (%)	Commit-ments (%)	No. of projects (%)	Commit-ments (%)
AFR	3	53					100	100
EAP	5	783					100	100
ECA	5	202	20	20			80	80
LCR	4	60					100	100
MNA	2	90					100	100
SAR	11	489					100	100
All Regions	**30**	**1,683**	**3**	**2**			**97**	**98**

Source: QAG.

TABLE D.8. RISK INDICATORS FOR ACTIVE AGRICULTURE SECTOR PROJECTS, BY REGION (JUNE 1999)

Risk indicator	Percentage of projects						
	AFR	EAP	ECA	LCR	MNA	SAR	All Regions
Number of projects	*66*	*55*	*38*	*46*	*28*	*35*	*268*
Effectiveness delays	11	2		20	18	3	9
Compliance with legal covenants	2	4	5	2		17	4
Management performance	9	13	5	2	14	9	9
Counterpart funds	18	11	8	13	11	11	13
Procurement progress	12	9	16	2	7	11	10
Financial performance	14	11	8	4	7	6	9
Environment/resettlement problems	3				14	3	3
Slow disbursements	30	16	11	20	25	26	22
History of past problems	11	7	3	7	11	14	9
Risky country	44	4	24	7	4	14	18
Risky subsector	11	5	18	2	4		7
Economic management	30	29	21	13	25	11	23
Golden rule	8		5	2			3

Source: QAG.

TABLE D.9. RISK INDICATORS FOR ACTIVE ENVIRONMENT SECTOR PROJECTS, BY REGION (JUNE 1999)

Risk indicator	Percentage of projects						
	AFR	EAP	ECA	LCR	MNA	SAR	All Regions
Number of projects	*16*	*20*	*15*	*29*	*6*	*10*	*96*
Effectiveness delays	13	5	7	17	33		11
Compliance with legal covenants				3		10	2
Management performance	13	5		10	17	20	9
Counterpart funds	6	10	20	21			13
Procurement progress	13		7		33	30	8
Financial performance	6	15	27				8
Environment/resettlement problems		5		3			2
Slow disbursements	25	20	20	31	50	20	26
History of past problems		5	7	14		10	7
Risky country	63		7	28	17	30	24
Risky subsector							
Economic management	25	30	40	21	17		24
Golden rule				3			1

Source: QAG.

TABLE D.10. RISK INDICATORS FOR ACTIVE FOREST-COMPONENT PROJECTS, BY REGION (JUNE 1999)

Risk indicator	Percentage of projects						
	AFR	EAP	ECA	LCR	MNA	SAR	All Regions
Number of projects	*17*	*24*	*6*	*20*	*5*	*8*	*80*
Effectiveness delays	18			25			10
Compliance with legal covenants	12			5		13	5
Management performance	6	21		15		13	13
Counterpart funds	6	17	17	25	20		15
Procurement progress	12	21				13	10
Financial performance	18	13	33				10
Environment/resettlement problems							
Slow disbursements	12	17	17	25		25	18
History of past problems		8		15			6
Risky country	41	4	33	30	20	38	25
Risky subsector	18	4					5
Economic management	29	38	17	25	40		28
Golden rule				5			1

Source: QAG.

TABLE D.11. RISK INDICATORS FOR ACTIVE FOREST PROJECTS, BY REGION (JUNE 1999)

Risk indicator	Percentage of projects						
	AFR	EAP	ECA	LCR	MNA	SAR	All Regions
Number of projects	*3*	*5*	*5*	*4*	*2*	*11*	*30*
Effectiveness delays				25	50		7
Compliance with legal covenants							
Management performance			20		50	9	10
Counterpart funds				25		9	7
Procurement progress			40	25			10
Financial performance				25			3
Environment/resettlement problems							
Slow disbursements	33	40	20			9	17
History of past problems	33					27	13
Risky country	67		20			9	13
Risky subsector							
Economic management	33	20	20			9	13
Golden rule	33						3

Source: QAG.

TABLE D.12. RISK INDICATORS FOR ACTIVE WORLD BANK PROJECTS, BY REGION (JUNE 1999)

Risk indicator	Percentage of projects						
	AFR	EAP	ECA	LCR	MNA	SAR	All Regions
Number of projects	*392*	*287*	*303*	*322*	*122*	*151*	*1,577*
Effectiveness delays	10	5	6	17	11	4	9
Compliance with legal covenants	6	4	4	5	7	10	6
Management performance	10	8	8	10	12	10	9
Counterpart funds	14	9	13	13	7	7	11
Procurement progress	11	11	10	4	10	15	10
Financial performance	12	11	12	6	7	11	10
Environment/resettlement problems	2	2	1	3	5	1	2
Slow disbursements	32	21	31	25	23	30	27
History of past problems	9	3	5	7	7	10	7
Risky country	49	2	16	18	7	19	21
Risky subsector	13	10	11	12	11	9	11
Economic management	38	34	36	16	26	10	29
Golden rule	3		2	3	2	1	2

Source: QAG.

ANNEX E: OPERATIONS EVALUATION DEPARTMENT PROJECT EVALUATIONS

Evaluation Criteria

The Operations Evaluation Department (OED) evaluates the development effectiveness of completed operations by examining their technical, financial, economic, social, and environmental aspects and rating their outcome, sustainability, and effect on institutional development. OED also evaluates the performance of the Bank, the borrower, and the implementing agencies. The evaluation process is based on the same criteria and policies that are used to judge new operations.

Outcome

A satisfactory outcome means that an operation has achieved most of its major goals efficiently. A project with an unsatisfactory rating may still provide significant benefits even though it failed to meet one or more of its major objectives. To evaluate outcome, completed operations are analyzed from three perspectives:

Relevance: Evaluators determine whether the goals of the operation were consistent with country and sectoral assistance strategies and whether the design was appropriate to one or more of the core World Bank goals of reducing poverty, protecting the environment, developing human resources, or fostering growth in the private sector.

Efficacy: Evaluators review the operation's effects and compare them with its goals, whether physical, financial, institutional, or policy related.

Efficiency: Evaluators assess outcomes in relation to inputs, looking at costs, implementation times, and economic and financial results. Where practicable, the evaluators re-estimate the economic rate of return in relation to a minimum threshold of 10 percent.

Sustainability

Sustainability is based on the probability—at the time of evaluation—that the achievements generated or expected to be generated in the operational plan will be sustained. To assess sustainability, evaluators examine borrower commitment; the policy environment; institutional/management effectiveness; and economic, social, technical, financial, institutional, and environmental viability. Sustainability differs from economic justification in that it focuses on features that contribute to the durability of the operation relative to the project's expected useful life and its likely resilience to external shocks and changing circumstances.

Institutional development impact

Institutional development impact is improvement in the ability of a country to use its human, organizational, and financial resources effectively. Evaluators assess the institutional development progress achieved, or expected, because of the operation. Not all operations have institutional development goals, but many depend on institutional change to achieve a lasting impact on development. Examples of such change include strengthening, eliminating, or reforming specific agencies; supporting regulatory or legal reforms; and supporting education and training.

Performance

OED analyzes Bank performance at each stage of the project cycle. It also considers several other dimensions of performance:

Borrower performance is evaluated in terms of the policy environment created for the project; the level of commitment of the government and of key institutions associated with the project; the provision and reliability of domestic funds for the operation; and the administrative procedures used and the quality of decisionmaking.

The performance of implementing agencies is rated on such elements as the quality of management and staff associated with the project; the use and effectiveness of technical assistance, including training, advisers, and contractual services; the adequacy of monitoring and evaluation systems; and the extent and quality of intended beneficiaries' participation, including their contribution to the project's outcome.

Exogenous factors include changes in prices and world market conditions; natural disasters; civil disorder and armed conflict; and actions of partners who are independent decisionmakers—for example, cofinanciers, NGOs, contractors, and suppliers.

OED Evaluations of Completed Bank Operations

OED evaluated a total of 1,590 Bank operations that exited the portfolio between 1992 and 1998. These projects had net commitments of $113.6 billion (1996 dollars). Figure E.1 shows that ratings on overall performance for the completed forest and forest-component projects are poorer than those for projects in the agriculture and environment sectors, and poorer than those for aggregate Bank projects. Moreover, the satisfactory ratings for outcome tend to be higher than

FIGURE E.1. SATISFACTORY OVERALL PERFORMANCE RATINGS, COMPLETED PROJECTS

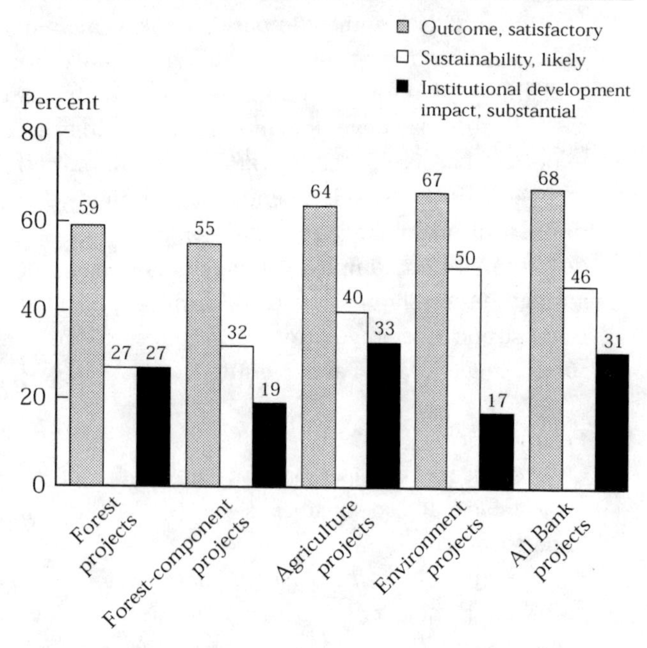

Source: OED data, 1992–98.

FIGURE E.2. SATISFACTORY BANK PERFORMANCE RATINGS, COMPLETED PROJECTS

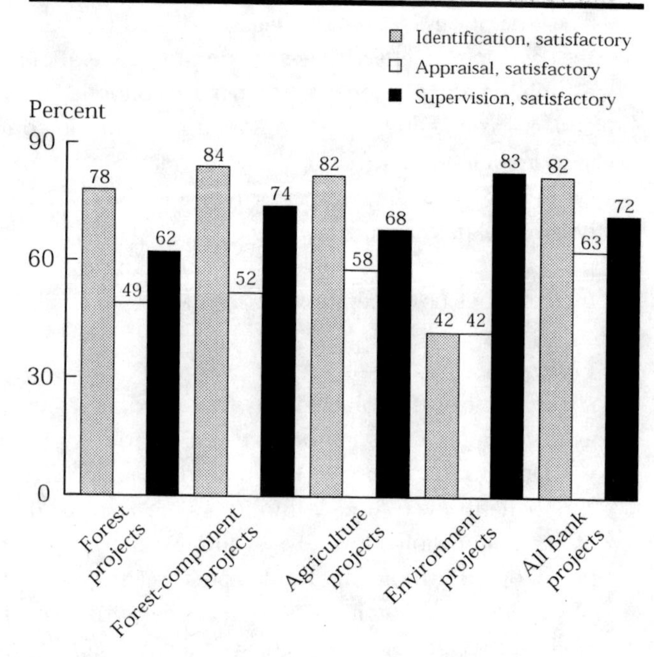

Source: OED data, 1992–98.

the likely ratings for sustainability. The substantial rating for institutional development is the lowest rating across all sectors.

Satisfactory Bank performance ratings for completed forest and forest-component projects are compared with those for projects in other relevant sectors in figure E.2. Across all sectors, "satisfactory" ratings for Bank performance at identification tend to be higher than the ratings during supervision. In contrast, appraisal ratings are the lowest. The ratings for the environment sector projects seem to be the lowest at identification and appraisal. The Bank performance in forest and forest-component projects is comparable to performance in the agriculture sector.

Borrower performance ratings tend to be lower than Bank performance ratings in all sectors. Interestingly, the ratings for completed forest-component projects are the highest for all stages, while forest project ratings are comparable to ratings for projects in the agriculture sector. Borrower preparation ratings are higher across all sectors, whereas implementation and compliance ratings are comparable (figure E.3).

FIGURE E.3. SATISFACTORY BORROWER PERFORMANCE RATINGS, COMPLETED PROJECTS

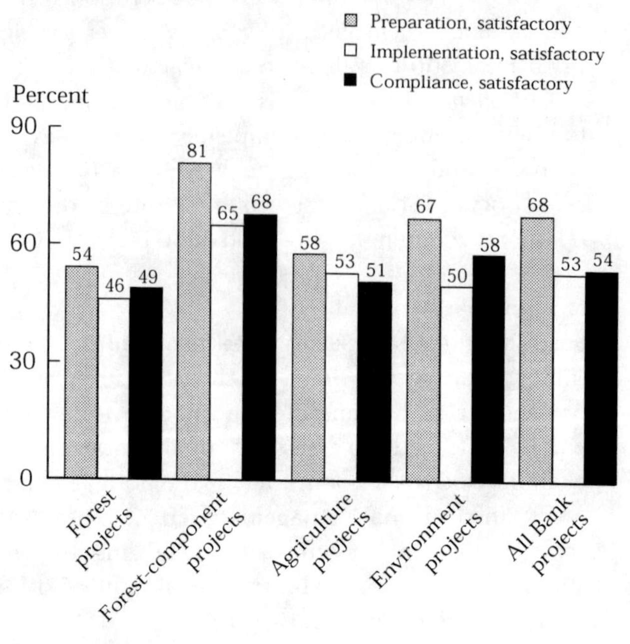

Source: OED data, 1992–98.

FIGURE E.4. SATISFACTORY OVERALL PERFORMANCE RATINGS, COMPLETED FOREST PROJECTS

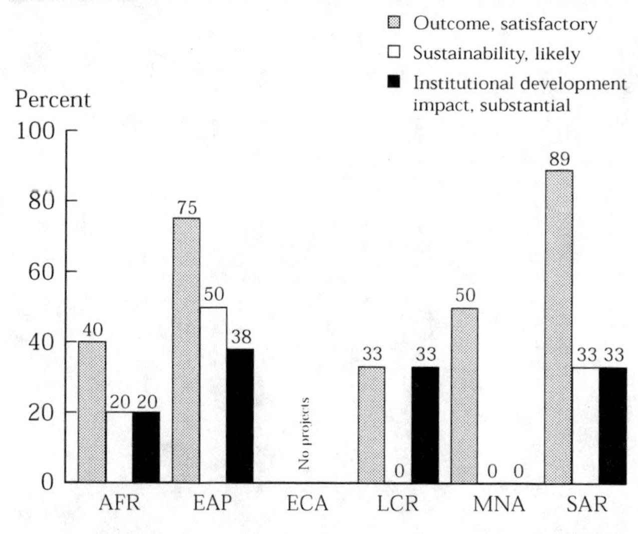

Source: OED data, 1992–98.

FIGURE E.5. SATISFACTORY BANK PERFORMANCE RATINGS, COMPLETED FOREST PROJECTS

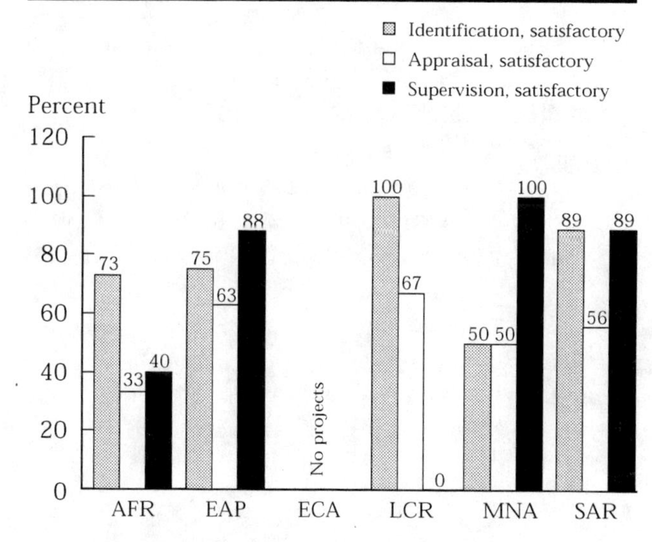

Source: OED data, 1992–98.

OED Evaluation of Completed Forest Projects

OED has evaluated 37 forest operations with net commitments of $1.0 billion that exited the portfolio during 1992–98. The regional distribution of the overall performance ratings (consisting of ratings for outcome, sustainability, and institutional development impact) shows that forest projects in the SAR and EAP Regions performed much better than those in the AFR or MNA Regions (figure E.4).

The Bank performance ratings at project identification and appraisal were also best in LCR, followed by SAR and EAP. However, satisfactory supervision ratings were lowest in the LCR. Bank performance was poorest overall in the AFR Region (figure E.5).

Borrower performance ratings were far lower than Bank performance ratings across all Regions. Borrower performance in the SAR and the EAP Regions was much better than in AFR or LCR (figure E.6).

The OED evaluation of 31 forest-component projects with net project commitments of $1.2 billion shows that projects in the EAP Region performed the best, and projects in SAR and AFR were the poorest

FIGURE E.6. SATISFACTORY BORROWER PERFORMANCE RATINGS, COMPLETED FOREST PROJECTS

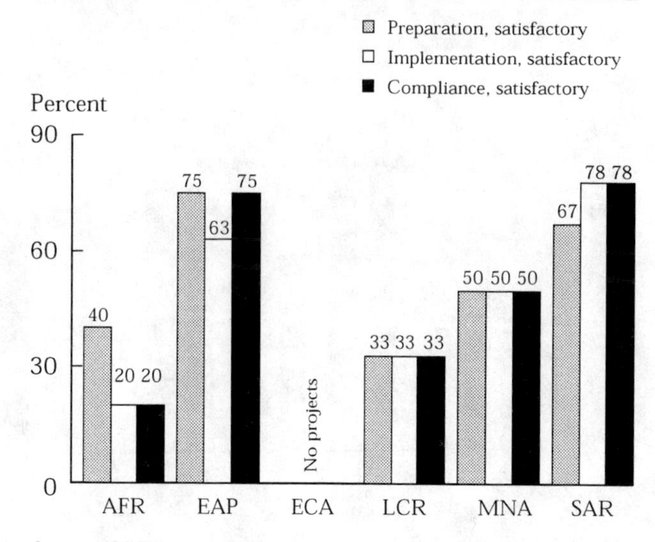

Source: OED data, 1992–98.

performers. As with forest projects, the satisfactory outcome ratings were much higher than the ratings for sustainability and institutional development impact across all Regions (figure E.7).

Bank performance for forest-component projects was best in the EAP Region, followed by SAR and LCR. In the AFR and MNA Regions Bank performance was not as good (figure E.8).

Borrower performance for forest-component projects was exceptionally good in the ECA and LCR, followed by the EAP Region. Borrower performance was weakest in SAR and AFR (figure E.9).

An overall comparison of OED ratings by Region for all Bank projects, for projects in the agriculture and environment sectors, for forest projects, and for forest-component projects is presented in tables E.1 through E.8.

FIGURE E.7. SATISFACTORY OVERALL PERFORMANCE RATINGS, COMPLETED FOREST-COMPONENT PROJECTS

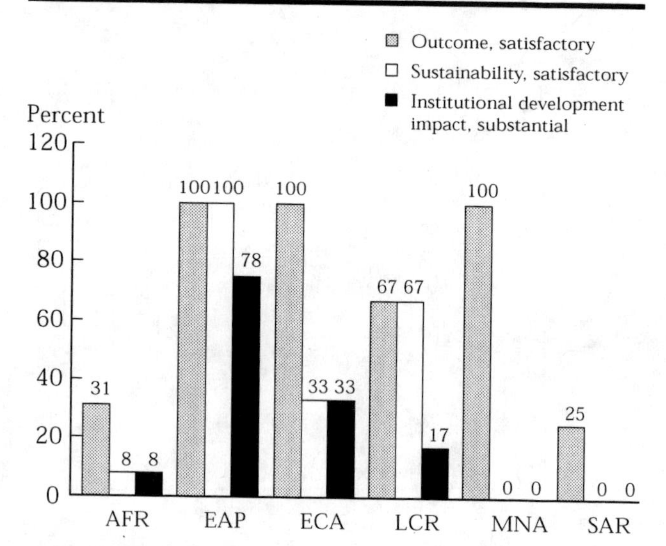

Source: OED data, 1992–98.

FIGURE E.8. SATISFACTORY BANK PERFORMANCE RATINGS, COMPLETED FOREST-COMPONENT PROJECTS

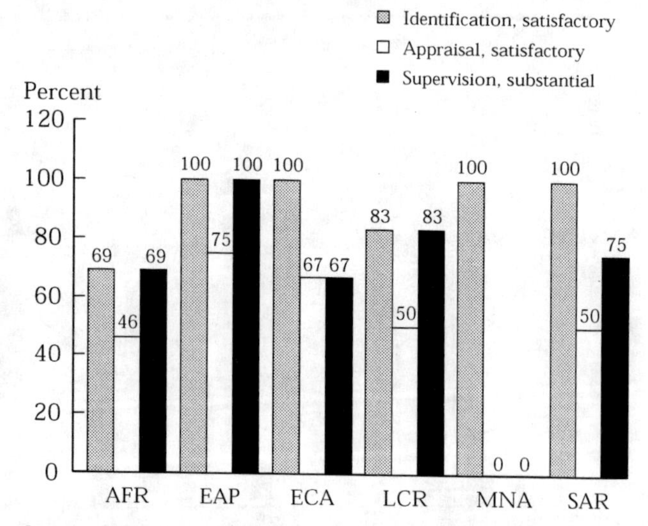

Source: OED data, 1992–98.

FIGURE E.9. SATISFACTORY BORROWER PERFORMANCE RATINGS, COMPLETED FOREST-COMPONENT PROJECTS

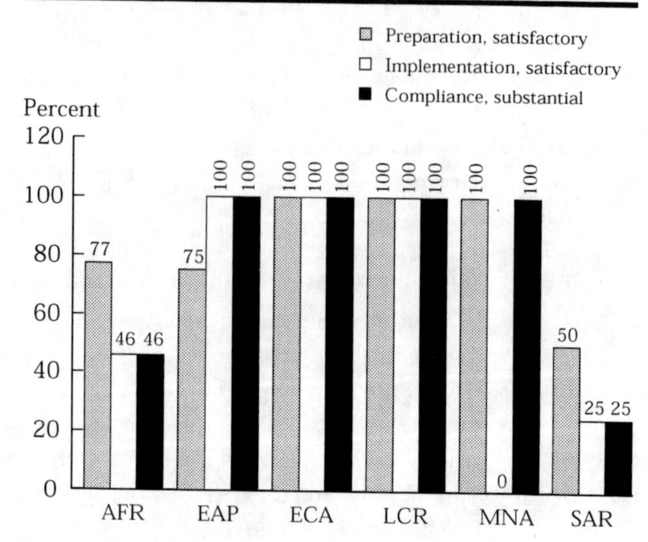

Source: OED data, 1992–98.

TABLE E.1. OVERALL PERFORMANCE RATINGS, COMPLETED FOREST PROJECTS RELATIVE TO PROJECTS IN OTHER SECTORS

Project	Evaluated projects		Outcome satisfactory		Sustainability likely		ID impact substantial	
	No. of projects	Commitments ($M, FY96)	No. of projects (%)	Commitments (%)	No. of projects (%)	Commitments (%)	No. of projects (%)	Commitments (%)
Forest projects	37	1,000	59	69	27	25	27	30
Forest-component projects	31	1,200	55	73	32	45	19	21
All agriculture projects	410	20,671	64	72	40	51	33	39
All environment projects	12	364	67	76	50	60	17	23
All World Bank projects	**1,590**	**113,592**	**68**	**76**	**46**	**56**	**31**	**36**

Source: OED data, 1992–98.

TABLE E.2. BANK PERFORMANCE RATINGS, COMPLETED FOREST PROJECTS RELATIVE TO PROJECTS IN OTHER SECTORS

Project	Evaluated projects		Identification satisfactory		Appraisal satisfactory		Supervision satisfactory	
	No. of projects	Commitments ($M FY96)	No. of projects (%)	Commitments (%)	No. of projects (%)	Commitments (%)	No. of projects (%)	Commitments (%)
Forest projects	37	1,000	78	86	49	45	62	66
Forest-component projects	31	1,200	84	93	52	46	74	72
All agriculture projects	410	20,671	82	80	58	58	68	71
All environment projects	12	364	42	39	42	39	83	89
All World Bank projects	**1,590**	**113,592**	**82**	**84**	**63**	**69**	**72**	**76**

Source: OED data, 1992–98.

TABLE E.3. BORROWER PERFORMANCE RATINGS, COMPLETED FOREST PROJECTS RELATIVE TO PROJECTS IN OTHER SECTORS

Project	Evaluated projects		Preparation satisfactory		Implementation satisfactory		Compliance satisfactory	
	No. of projects	Commitments ($M FY96)	No. of projects (%)	Commitments (%)	No. of projects (%)	Commitments (%)	No. of projects (%)	Commitments (%)
Forest projects	37	1,000	54	68	46	47	49	56
Forest-component projects	31	1,200	81	83	65	74	68	80
All agriculture projects	410	20,671	58	60	53	63	51	61
All environment projects	12	364	67	68	50	48	58	70
All World Bank projects	**1,590**	**113,592**	**68**	**74**	**53**	**63**	**54**	**63**

Source: OED data, 1992–98.

TABLE E.4. OVERALL PERFORMANCE RATINGS, COMPLETED AGRICULTURE SECTOR PROJECTS, BY REGION

Region	Evaluated projects		Outcome satisfactory		Sustainability likely		ID impact substantial	
	No. of projects	Commitments ($M FY96)	No. of projects (%)	Commitments (%)	No. of projects (%)	Commitments (%)	No. of projects (%)	Commitments (%)
AFR	146	4,020	53	62	32	32	27	28
EAP	65	4,032	83	86	66	67	45	40
ECA	24	2,067	83	92	71	75	54	65
LCR	55	4,712	64	72	44	62	38	41
MNA	43	1,739	60	65	28	45	30	36
SAR	77	4,102	68	63	26	31	27	33
All Regions	**410**	**20,671**	**64**	**72**	**40**	**51**	**33**	**39**

Source: OED data, 1992–98.
Note: ID = institutional development.

TABLE E.5. OVERALL PERFORMANCE RATINGS, COMPLETED ENVIRONMENT SECTOR PROJECTS, BY REGION

Region	Evaluated projects		Outcome satisfactory		Sustainability likely		ID impact substantial	
	No. of projects	Commitments ($M FY96)	No. of projects (%)	Commitments (%)	No. of projects (%)	Commitments (%)	No. of projects (%)	Commitments (%)
AFR	3	72	33	35				
EAP	2	50	50	60	50	60		
ECA	1	19	100	100	100	100	100	100
LCR	6	224	83	91	67	76	17	30
MNA	—	—	—	—	—	—	—	—
SAR	—	—	—	—	—	—	—	—
All Regions	**12**	**364**	**67**	**76**	**50**	**60**	**17**	**23**

Source: OED data, 1992–98.
Note: ID = institutional development.

TABLE E.6. OVERALL PERFORMANCE RATINGS, COMPLETED WORLD BANK PROJECTS, BY REGION

Region	Evaluated projects		Outcome satisfactory		Sustainability likely		ID impact substantial	
	No. of projects	Commitments ($M FY96)	No. of projects (%)	Commitments (%)	No. of projects (%)	Commitments (%)	No. of projects (%)	Commitments (%)
AFR	530	17,607	55	60	29	27	23	23
EAP	274	23,847	84	87	69	73	40	42
ECA	116	13,788	76	80	59	65	42	44
LCR	291	28,708	73	80	53	63	38	42
MNA	152	9,279	66	66	46	39	28	26
SAR	227	20,362	67	71	42	55	28	32
All Regions	**1,590**	**113,592**	**68**	**76**	**46**	**56**	**31**	**36**

Source: OED data, 1992–98.
Note: ID = institutional development.

TABLE E.7. OVERALL PERFORMANCE RATINGS, COMPLETED FOREST-COMPONENT PROJECTS, BY REGION

Region	Evaluated projects		Outcome satisfactory		Sustainability likely		ID impact substantial	
	No. of projects	Commitments ($M FY96)	No. of projects (%)	Commitments (%)	No. of projects (%)	Commitments (%)	No. of projects (%)	Commitments (%)
AFR	13	258	31	21	8	6	8	4
EAP	4	272	100	100	100	100	75	58
ECA	3	253	100	100	33	10	33	10
LCR	6	258	67	84	67	90	17	26
MNA	1	50	100	100	—	—	—	—
SAR	4	108	25	26	—	—	—	—
All Regions	**31**	**1,200**	**55**	**73**	**32**	**45**	**19**	**21**

Source: OED data, 1992–98.
Note: ID = institutional development.

TABLE E.8. OVERALL PERFORMANCE RATINGS, COMPLETED FOREST PROJECTS, BY REGION

Region	Evaluated projects		Outcome satisfactory		Sustainability likely		ID impact substantial	
	No. of projects	Commitments ($M FY96)	No. of projects (%)	Commitments (%)	No. of projects (%)	Commitments (%)	No. of projects (%)	Commitments (%)
AFR	15	391	40	55	20	9	20	38
EAP	8	209	75	79	50	37	38	33
ECA	—	—	—	—	—	—	—	—
LCR	3	82	33	48	—	—	33	48
MNA	2	40	50	47	—	—	—	—
SAR	9	277	89	92	33	49	33	15
All Regions	**37**	**1,000**	**59**	**69**	**27**	**25**	**27**	**30**

Source: OED data, 1992–98.
Note: ID = institutional development.

TABLE E.9. BANK PERFORMANCE RATINGS, COMPLETED AGRICULTURE SECTOR PROJECTS, BY REGION

Region	Evaluated projects		Identification satisfactory		Appraisal satisfactory		Supervision satisfactory	
	No. of projects	Commitments ($M FY96)	No. of projects (%)	Commitments (%)	No. of projects (%)	Commitments (%)	No. of projects (%)	Commitments (%)
AFR	146	4,020	82	82	58	60	64	69
EAP	65	4,032	92	96	71	64	78	74
ECA	24	2,067	83	90	71	71	79	91
LCR	55	4,712	73	63	53	51	58	69
MNA	43	1,739	79	69	53	39	63	60
SAR	77	4,102	83	82	52	62	71	69
All Regions	**410**	**20,671**	**82**	**80**	**58**	**58**	**68**	**71**

Source: OED data, 1992–98.

TABLE E.10. BANK PERFORMANCE RATINGS, COMPLETED ENVIRONMENT SECTOR PROJECTS, BY REGION

	Evaluated projects		Identification satisfactory		Appraisal satisfactory		Supervision satisfactory	
Region	No. of projects	Commit-ments ($M FY96)	No. of projects (%)	Commit-ments (%)	No. of projects (%)	Commit-ments (%)	No. of projects (%)	Commit-ments (%)
AFR	3	72	33	35	33	35	67	72
EAP	2	50	50	60	50	60	50	60
ECA	1	19	100	100	100	100	100	100
LCR	6	224	33	31	33	31	100	100
MNA	—	—	—	—	—	—	—	—
SAR	—	—	—	—	—	—	—	—
All Regions	**12**	**364**	**42**	**39**	**42**	**39**	**83**	**89**

Source: OED data, 1992–98.

TABLE E.11. BANK PERFORMANCE RATINGS, ALL WORLD BANK PROJECTS, BY REGION

	Evaluated projects		Identification satisfactory		Appraisal satisfactory		Supervision satisfactory	
Region	No. of projects	Commit-ments ($M FY96)	No. of projects (%)	Commit-ments (%)	No. of projects (%)	Commit-ments (%)	No. of projects (%)	Commit-ments (%)
AFR	530	17,607	78	80	58	63	67	71
EAP	274	23,847	89	91	76	78	80	80
ECA	116	13,788	87	89	76	80	82	89
LCR	291	28,708	83	82	61	65	70	71
MNA	152	9,279	81	77	61	57	71	71
SAR	227	20,362	82	81	56	65	74	78
All Regions	**1,590**	**113,592**	**82**	**84**	**63**	**69**	**72**	**76**

Source: OED data, 1992–98.

TABLE E.12. BANK PERFORMANCE RATINGS, COMPLETED FOREST-COMPONENT PROJECTS, BY REGION

	Evaluated projects		Identification satisfactory		Appraisal satisfactory		Supervision satisfactory	
Region	No. of projects	Commit-ments ($M FY96)	No. of projects (%)	Commit-ments (%)	No. of projects (%)	Commit-ments (%)	No. of projects (%)	Commit-ments (%)
AFR	13	258	69	75	46	42	69	76
EAP	4	272	100	100	75	58	100	100
ECA	3	253	100	100	67	28	67	28
LCR	6	258	83	92	50	58	83	91
MNA	1	50	100	100	—	—	—	—
SAR	4	108	100	100	50	56	75	76
All Regions	**31**	**1,200**	**84**	**93**	**52**	**46**	**74**	**72**

Source: OED data, 1992–98.

TABLE E.13. BANK PERFORMANCE RATINGS, COMPLETED FOREST PROJECTS, BY REGION

Region	Evaluated projects		Identification satisfactory		Appraisal satisfactory		Supervision satisfactory	
	No. of projects	Commit-ments ($M FY96)	No. of projects (%)	Commit-ments (%)	No. of projects (%)	Commit-ments (%)	No. of projects (%)	Commit-ments (%)
AFR	15	391	73	82	33	37	40	52
EAP	8	209	75	79	63	50	88	84
ECA	—	—	—	—	—	—	—	—
LCR	3	82	100	100	67	80	—	—
MNA	2	40	50	47	50	47	100	100
SAR	9	277	89	97	56	42	89	87
All Regions	**37**	**1,000**	**78**	**86**	**49**	**45**	**62**	**66**

Source: OED data, 1992–98.

TABLE E.14. BORROWER PERFORMANCE RATINGS, COMPLETED AGRICULTURE SECTOR PROJECTS, BY REGION

Region	Evaluated projects		Preparation satisfactory		Implementation satisfactory		Compliance satisfactory	
	No. of projects	Commit-ments ($M FY96)	No. of projects (%)	Commit-ments (%)	No. of projects (%)	Commit-ments (%)	No. of projects (%)	Commit-ments (%)
AFR	146	4,020	53	62	32	32	27	28
EAP	65	4,032	72	72	77	81	75	73
ECA	24	2,067	71	63	71	77	79	87
LCR	55	4,712	42	44	58	71	60	67
MNA	43	1,739	72	55	53	53	60	76
SAR	77	4,102	56	63	62	64	57	54
All Regions	**410**	**20,671**	**58**	**60**	**53**	**63**	**51**	**61**

Source: OED data, 1992–98.

TABLE E.15. BORROWER PERFORMANCE RATINGS, COMPLETED ENVIRONMENT SECTOR PROJECTS, BY REGION

Region	Evaluated projects		Preparation satisfactory		Implementation satisfactory		Compliance satisfactory	
	No. of projects	Commit-ments ($M FY96)	No. of projects (%)	Commit-ments (%)	No. of projects (%)	Commit-ments (%)	No. of projects (%)	Commit-ments (%)
AFR	3	72	33	35				
EAP	2	50	50	60	50	60	50	60
ECA	1	19	100	100	100	100	—	—
LCR	6	224	83	78	67	56	100	100
MNA	—	—	—	—	—	—	—	—
SAR	—	—	—	—	—	—	—	—
All Regions	**12**	**364**	**67**	**68**	**50**	**48**	**58**	**70**

Source: OED data, 1992–98.

TABLE E.16. BORROWER PERFORMANCE RATINGS, COMPLETED WORLD BANK PROJECTS, BY REGION

Region	Evaluated projects		Preparation satisfactory		Implementation satisfactory		Compliance satisfactory	
	No. of projects	Commitments ($M FY96)	No. of projects (%)	Commitments (%)	No. of projects (%)	Commitments (%)	No. of projects (%)	Commitments (%)
AFR	530	17,607	55	60	29	27	23	23
EAP	274	23,847	83	88	76	79	77	78
ECA	116	13,788	78	75	68	72	72	69
LCR	291	28,708	68	71	59	62	68	68
MNA	152	9,279	72	66	61	56	64	65
SAR	227	20,362	68	79	62	71	65	68
All Regions	**1,590**	**113,592**	**68**	**74**	**53**	**63**	**54**	**63**

Source: OED data, 1992–98.

TABLE E.17. BORROWER PERFORMANCE RATINGS, COMPLETED FOREST-COMPONENT PROJECTS, BY REGION

Region	Evaluated projects		Preparation satisfactory		Implementation satisfactory		Compliance satisfactory	
	No. of projects	Commitments ($M FY96)	No. of projects (%)	Commitments (%)	No. of projects (%)	Commitments (%)	No. of projects (%)	Commitments (%)
AFR	13	258	77	86	46	31	46	38
EAP	4	272	75	58	100	100	100	100
ECA	3	253	100	100	100	100	100	100
LCR	6	258	100	100	100	100	100	100
MNA	1	50	100	100	—	—	100	100
SAR	4	108	50	56	25	26	25	26
All Regions	**31**	**1,200**	**81**	**83**	**65**	**74**	**68**	**80**

Source: OED data, 1992–98.

TABLE E.18. BORROWER PERFORMANCE RATINGS, COMPLETED FOREST PROJECTS, BY REGION

Region	Evaluated projects		Preparation satisfactory		Implementation satisfactory		Compliance satisfactory	
	No. of projects	Commitments ($M FY96)	No. of projects (%)	Commitments (%)	No. of projects (%)	Commitments (%)	No. of projects (%)	Commitments (%)
AFR	15	391	40	55	20	9	20	38
EAP	8	209	75	92	63	78	75	92
ECA	—	—	—	—	—	—	—	—
LCR	3	82	33	48	33	48	33	48
MNA	2	40	50	47	50	47	50	47
SAR	9	277	67	76	78	79	78	58
All Regions	**37**	**1,000**	**54**	**68**	**46**	**47**	**49**	**56**

Source: OED data, 1992–98.

ANNEX F: COUNTRY CASE STUDY DATA

TABLE F.1. WORLD BANK LENDING OPERATIONS IN THE CASE STUDY COUNTRIES, 1984–91

Country/ category	Total lending		Adjustment lending		Investment lending		Agricultural lending		Environment lending		Forestry lending		Forest-component lending	
	No. of projects	Commit-ments ($M)	No. of projects	Commit-ments ($M)	No. of projects	Commit-ments ($M)	No. of projects	Commit-ments ($M)	No. of projects	Commit-ments ($M)	No. of projects	Commit-ments ($M)	No. of projects	Commit-ments ($M)
World Bank lending	1,919	148,946	265	34,611	1,654	114,336	464	29,282	15	889	41	1,682	32	291
Brazil	72	10,599	3	1,155	69	9,444	28	3,540	4	263	1	49	2	125
Cameroon	14	733	1	150	13	583	7	232	—	—	—	—	1	1
China	85	9,913	1	300	84	9,613	25	3,024	—	—	3	404	2	6
Costa Rica	5	270	2	180	3	90	1	26	—	—	—	—	—	—
India	97	20,242	—	—	97	20,242	29	4,544	4	459	3	224	3	17
Indonesia	79	10,482	4	1,200	75	9,282	20	1,642	—	—	2	54	—	—
Total country lending	**352**	**52,239**	**11**	**2,985**	**341**	**49,254**	**110**	**13,008**	**8**	**722**	**9**	**731**	**8**	**149**
Percent, overall Bank lending	**18**	**35**	**4**	**9**	**21**	**43**	**24**	**44**	**53**	**81**	**22**	**43**	**25**	**51**

Source: World Bank databases.

TABLE F.2. WORLD BANK LENDING OPERATIONS IN THE CASE STUDY COUNTRIES, 1992–99

Country/ category	Total lending		Adjustment lending		Investment lending		Agricultural lending		Environment lending		Forestry lending		Forest-component lending	
	No. of projects	Commit-ments ($M)	No. of projects	Commit-ments ($M)	No. of projects	Commit-ments ($M)	No. of projects	Commit-ments ($M)	No. of projects	Commit-ments ($M)	No. of projects	Commit-ments ($M)	No. of projects	Commit-ments ($M)
World Bank lending	2,175	189,336	378	57,092	1,797	132,243	336	22,912	106	5,899	34	1,722	94	1,790
Brazil	66	9,349	3	1,760	63	7,589	15	978	8	936	—	—	4	231
Cameroon	18	817	10	626	8	192	2	38	—	—	—	—	—	—
China	125	22,066	—	—	125	22,066	28	5,404	13	1,707	3	550	15	305
Costa Rica	5	212	1	100	4	112	1	41	—	—	—	—	—	—
India	80	15,202	4	1,450	76	13,752	21	2,661	3	360	8	460	3	179
Indonesia	84	10,742	4	2,400	80	8,342	14	1,208	4	198	—	—	7	88
Total country lending	**378**	**58,389**	**22**	**6,336**	**356**	**52,053**	**81**	**10,330**	**28**	**3,201**	**11**	**1,010**	**29**	**803**
Percent, overall Bank lending	**17**	**31**	**6**	**11**	**20**	**39**	**24**	**45**	**26**	**54**	**32**	**59**	**31**	**45**

Source: World Bank databases.

TABLE F.3. PERCENTAGE CHANGE IN WORLD BANK LENDING OPERATIONS IN CASE STUDY COUNTRIES AFTER 1991

Country/ category	Total lending		Adjustment lending		Investment lending		Agricultural lending		Environment lending		Forestry lending		Forest-component lending	
	No. of projects	Commit-ments ($M)	No. of projects	Commit-ments ($M)	No. of projects	Commit-ments ($M)	No. of projects	Commit-ments ($M)	No. of projects	Commit-ments ($M)	No. of projects	Commit-ments ($M)	No. of projects	Commit-ments ($M)
World Bank lending	13	27	43	65	9	16	-28	-22	607	564	-17	2	194	515
Brazil	-8	-12	—	52	-9	-20	-46	-72	100	256	-100	-100	100	85
Cameroon	29	11	900	317	-38	-67	-71	-84	—	—	—	—	-100	-100
China	47	123	-100	-100	49	130	12	79.	—	—	—	36	650	5,018
Costa Rica	—	-21	-50	-44	33	25	—	58	—	—	—	—	—	—
India	-18	-25	—	—	-22	-32	-28	-41	-25	-22	167	106	—	958
Indonesia	6	2	—	100	7	-10	-30	-26	—	—	-100	-100	—	—
Total country lending	**7**	**12**	**100**	**112**	**4**	**6**	**-26**	**-21**	**250**	**344**	**22**	**38**	**263**	**440**
Percent, overall Bank lending	**13**	**27**	**43**	**65**	**9**	**16**	**-28**	**-22**	**607**	**564**	**-17**	**2**	**194**	**515**

Source: World Bank databases.

ANNEX G: THE SURVEYS

As part of the consultative process for this review, OED sought input from members of the World Bank's Forestry Community of Practice as well as from participants in two of the Bank's new initiatives—the CEO Forum and the WB/WWF Alliance—designed to enhance the conservation and sustainable management of the world's forests. A summary of conclusions from the survey results is incorporated in Chapter 4 of this report. This annex provides background information and aggregate findings from the three independently conducted surveys.

The Staff Survey

OED held a series of focus group sessions in late winter/early spring 1999 to gain insights into specific issues as seen by country and sector managers, lead macro and sector economists, and task managers and specialists involved in addressing issues related to the forest sector. When the sessions ended, results of the meetings were compiled and prepared by the facilitator, Madelyn Blair of Peleri, Inc. The ideas expressed by some participants in the sessions required statistical validation. So in July 1999 OED sent a questionnaire based on the preliminary focus group findings to 100 Bank staff who belong to the Bank's Forestry Community of Practice. OED received 40 responses, or a 40 percent response rate.

The CEO Forum Questionnaire

Of the 31 industry, NGO, and government ministry representatives who belong to the CEO Forum (see box 4.1 in the main report), to all of whom survey questionnaires were sent, OED received 15 responses—for a 48 percent response rate.

The WB/WWF Alliance Questionnaire

A WB/WWF Alliance questionnaire was sent to 47 World Bank and 46 WWF staff members; OED received 20 responses from Bank staff (for a 43 percent response rate) and 14 from WWF staff (for a 30 percent response rate). The overall response rate was 37 percent.

In many cases, Bank staff working on WB/WWF Alliance issues are located within the Forestry Community of Practice, so the Bank Staff Survey and WB/WWF Alliance responses reflect some overlap of opinion, (seven Bank staff responded to both the WB/WWF Alliance Survey and the Staff Survey). Staff members who were gracious enough to return both sets of questions gave OED feedback on policy questions and also gave their opinion about the objectives and progress of the alliance. All respondents surveyed were assured anonymity.

Survey results follow.

THE OED STAFF SURVEY

Sample Size:	**100**
Responses:	**40**
Response Rate:	**40%**
Date Sent:	**July 1999**
Received:	**July-September 1999**

I. General Information

(Some staff provided more than one response per question.)

Region(s) of expertise

AFR	12
EAP	12
ECA	10
LCR	13
MNA	10
SAR	3
Blank	4

Title

Environment (specialist, economist, adviser)	9
Natural resources management (economist, specialist)	8
Forestry (specialist, adviser, officer)	6
Manager (sector, knowledge, task)	7
Biodiversity specialist	2
Operations analyst	2
Communications specialist	1
Consultant – economics and financial analysis	1
Ecologist	1
Socio-economist	1
Swiss Secondment	1
No answer	5

Network affiliation

Primary	
Environmentally and Socially Sustainable Development Network	92.5% (37)
Blank	7.5% (3)
Secondary	
Poverty Reduction and Economic Management Network	5% (2)
Operational Core Services Network	2.5% (1)
Blank	92.5% (37)

Family affiliation

Primary	
Environment Department	45% (18)
Rural Development Department	25% (10)
Social Development Department	10% (4)
Blank	20% (8)
Secondary	
Environment Department	17.5% (7)
Rural Development Department	25% (10)
Social Development Department	7.5% (3)
Environment Department/ Social Development Department	2.5% (1)
Energy, Mining, and Telecommunications Department	2.5% (1)
Blank	45% (18)

Do you have any experience working outside the Bank Group in the past 10 years?

Yes	77.5% (31)
No	20% (8)
Blank	2.5% (1)

What is your disciplinary background?

Economics (business, natural resources management, forest, and political economist)	15
Forestry (technical specialist, ecologist, policy and science)	12
NRM (economist, socio-economist)	4
Ecology (ecologist, tropical ecologist)	3
Agriculture (agriculturist, agricultural engineer)	3
Anthropologist	3
Environment (environmental scientist, environmental technical specialist)	2
Energy (renewable energy engineer, scientist)	2
Biodiversity (technical specialist)	2
Communication	1
Veterinarian	1

Please indicate the type of forestry-related Bank experience you have.

Project preparation	31
Project supervision	28
Economic and sector work	18
Participation in CAS preparation	15
Policy dialogue with borrower(s)	21
Other[a]	15

a. Regional forest policy development, environmental assessment review, environmental action plan, BAP, training portfolio analysis, knowledge management and research, NRM strategy, donor coordination, GEF work, WB/WWF Alliance.

For the country or countries you currently work on, are forestry-related issues adequately reflected in the CAS?

Yes	25% (10)
No	42.5% (17)
Partially	7.5% (3)
Blank	25% (10)

Do you think adequate analytical work-for example, economic and sector work-underpins the Bank's operations?

Yes	50% (20)
No	22.5% (9)
Varies by country	5% (2)
Blank	22.5% (9)

Do you think adequate analytical work-for example, economic and sector work-underpins the Bank's forest policy dialogue?

Yes	27.5% (11)
No	42.5% (17)
Varies by country	7.5% (3)
Blank	22.5% (9)

II. The Bank's 1991 Forest Strategy

(a) the prevention of excessive rates of deforestation by expanding efforts toward the conservation, protection, and management of the world's remaining forests and woodlands, especially tropical moist forests and

(b) to ensure adequate planting of new trees to meet the rapidly growing demand for fuelwood, fodder, building poles, and other products and to ensure that adequate tree cover remains in rural areas for protection of soil and water resources.

Do you agree or disagree with the two most crucial challenges facing the forest sector as stated in the World Bank's 1991 Forest Strategy?

	SA	A	D	SD	NA
i. The thrust of statement (a) quoted above	35% (14)	47.5% (19)	10% (4)	0	7.5% (3)
ii. The thrust of statement (b) quoted above	20% (8)	57.5% (23)	15% (6)	0	7.5% (3)
iii. The emphasis on the conservation of tropical moist forests	17.5% (7)	45% (18)	27.5% (11)	2.5% (1)	7.5% (3)
iv. The Bank has contributed toward slowing down the rates of deforestation	0	20% (8)	50% (20)	15% (6)	15% (6)

SA: Strongly Agree; A: Agree; D: Disagree; SD: Strongly Disagree; NA: Not Available.

How do you rate the Bank's performance with respect to following?

	HS	S	U	HU	NA
i. Promoting planting of new trees in client countries	0	30% (12)	50% (20)	7.5% (3)	12.5% (5)
ii. Protection of natural forests	0	32.5% (13)	47.5% (19)	10% (4)	10% (4)
iii. Policy reforms that impact on forests	0	42.5% (17)	37.5% (15)	10% (4)	10% (4)
iv. Institutional reforms that impact on forests	0	32.5% (13)	47.5% (19)	12.5% (5)	7.5% (3)
v. Multisectoral approach to forest development	2.5% (1)	27.5% (11)	45% (18)	17.5% (7)	7.5% (3)
vi. Borrower capacity building	0	40% (16)	45% (18)	7.5% (3)	7.5% (3)
vii. Consideration of forestry-poverty interactions	2.5% (1)	27.5% (11)	37.5% (15)	20% (8)	12.5% (5)

HS: Highly Satisfactory; S: Satisfactory; U: Unsatisfactory; HU: Highly Unsatisfactory; NA: Not Available.

III. Bank Financing of Commercial Logging

The Bank does not finance commercial logging operations or the purchase of logging equipment for use in primary tropical moist forest. In borrowing countries where logging is being done in such forests, the Bank seeks the government's commitment to move toward sustainable management of those forests and to retain as much effective forest cover as possible. Where the government has made this commitment, the Bank may finance improvements in the planning, monitoring, and field control of forestry operations to maximize the capability of responsible agencies to carry out the sustainable management of the resource.

Do you agree or disagree with the following issues regarding OP 4.36, which were raised by some of you during the focus group discussions?

	SA	A	D	SD	NA
i. The policy of not financing commercial logging is irrelevant as it has not affected the rate of deforestation in client countries.	7.5% (3)	62.5% (25)	22.5% (9)	5% (2)	2.5% (1)
ii. The OP provides sufficient flexibility to address the key issues related to the logging of primary tropical moist forests.	2.5% (1)	22.5% (9)	57.5% (23)	7.5% (3)	10% (4)
iii. The OP has contributed to meeting the Bank's objective of sustainable forest management.	0	12.5% (5)	65% (26)	5% (2)	17.5% (7)
iv. Bank operations and policy dialogue since 1991 have helped reduce logging activities in client countries.	2.5% (1)	10% (4)	45% (18)	32.5% (13)	10% (4)

SA: Strongly Agree; A: Agree; D: Disagree; SD: Strongly Disagree; NA: Not Available.

IV. Internal Constraints on Strategy Implementation

Please indicate if you agree or disagree with the following statements about constraints within the Bank.

	SA	A	D	SD	NA
i. Relative to other sectors, country managers perceive forest projects to entail higher "transactions cost" but lower payoffs (i.e, smaller size, slower disbursement rates, etc.).	32.5% (13)	45% (18)	5% (2)	0	17.5% (7)
ii. This perception has led to lower operations than would otherwise have been the case.	35% (14)	47.5% (19)	7.5% (3)	0	10% (4)
iii. The current forest policy has contributed to these perceptions.	17.5% (7)	37.5% (15)	30% (12)	0	15% (6)
iv. Internal budgetary resources are insufficient to conduct high-quality economic and sector work.	35% (14)	37.5% (15)	15% (6)	2.5% (1)	10% (4)
v. Internal human capacity is insufficient to carry out adequate economic and sector work.	30% (12)	47.5% (19)	12.5% (5)	2.5% (1)	7.5% (3)
vi. Past projects have performed poorly.	7.5% (3)	37.5% (15)	25% (10)	0	30% (12)
vii. The complexity of the design of lending operations given their inherent multi-disciplinary and multisectoral nature.	7.5% (3)	47.5% (19)	15% (6)	2.5% (1)	27.5% (11)
viii. The task managers are not adequately qualified to handle forest-related operations.	12.5% (5)	27.5% (11)	45% (18)	0	15% (6)
ix. The forest sector staff have insufficient voice vis-à-vis the country managers.	22.5% (9)	47.5% (19)	12.5% (5)	0	17.5% (7)
x. Poor matrix management.	20% (8)	37.5% (15)	10% (4)	0	32.5% (13)
xi. The resources for project preparation are inadequate.	20% (8)	30% (12)	27.5% (11)	0	22.5% (9)

SA: Strongly Agree; A: Agree; D: Disagree; SD: Strongly Disagree; NA: Not Available.

V. External Constraints to Strategy Implementation

Please indicate if you agree or disagree with the following constraints that are external to the Bank.

	SA	A	D	SD	NA
i. Corruption in implementing agencies	35% (14)	40% (16)	17.5% (17)	0	7.5% (3)
ii. Inadequate appreciation of key issues by policymakers	17.5% (7)	57.5% (23)	20% (8)	0	5% (2)
iii. Insufficient implementation capacity	32.5% (13)	55% (22)	5% (2)	5% (2)	2.5% (1)
iv. Insufficient voice of forestry ministry vis-à-vis finance ministries	20% (8)	50% (20)	15% (6)	10% (4)	5% (2)
v. Availability of cheaper and more flexible sources of funds	12.5% (5)	27.5% (11)	40% (16)	2.5% (1)	17.5% (7)
vi. Controversial nature of the forest-related policies resulting in high levels of scrutiny by nongovernmental organizations	15% (6)	52.5% (21)	20% (8)	2.5% (1)	10% (4)

SA: Strongly agree; A: Agree; D: Disagree; SD: Strongly Disagree; Not Available.

VI. Network Leadership

Based on your implementation experience of the 1991 Forest Strategy, where does the leadership on the following aspects of forest sector operations come from?

	Rural Development Department	Environment Department	None	Other	NA
i. Intellectual leadership	15% (6)	10% (4)	25% (10)	22.5% (9)	27.5% (11)
ii. Operational support/cross support	25% (10)	15% (6)	10% (4)	25% (10)	25% (10)
iii. Innovative ideas	10% (4)	25% (10)	10% (4)	20% (8)	35% (14)
iv. Resources to operationalize innovative ideas	2.5% (1)	20% (8)	20% (8)	20% (8)	37.5% (15)
v. Quality control	10% (4)	12.5% (5)	32.5% (13)	15% (6)	30% (12)

Please rate the following with respect to the Network leadership.

	HS	S	U	HU	NA
i. The division of Forest Sector Management between ENV and RDV	0	10% (4)	32.5% (13)	30% (12)	27.5% (11)
ii. Intellectual leadership on Forest Sector Issues	0	30% (12)	32.5% (13)	12.5% (5)	25% (10)
iii. Knowledge management/dissemination of best practices	0	37.5% (15)	37.5% (15)	7.5% (3)	17.5% (7)
iv. Operational support/cross support	2.5% (1)	25% (10)	30% (12)	2.5% (1)	40% (16)
v. Leadership for operationalizing innovative ideas	0	15% (6)	47.5% (19)	7.5% (3)	30% (12)
vi. Resources for operationalizing innovative ideas	0	12.5% (5)	45% (18)	15% (6)	27.5% (11)
vii. Peer review process	2.5% (1)	25% (10)	37.5% (15)	5% (2)	30% (12)

HS: Highly Satisfactory; S: Satisfactory; U: Unsatisfactory; HU: Highly Unsatisfactory; NA: Not Available.

Which Family do you think should have primary responsibility for the forest sector?

RDV	32.5% (13)	Cross-cutting matrix	7.5% (3)
ENV	10% (4)	Create a separate family to govern NRM	2.5% (1)
SDV	2.5% (1)	Doesn't matter	5% (2)
RDV/ENV	12.5% (5)	No answer	17.5% (7)
RDV/ENV/SDV	10% (4)		

VII. Bank Leadership in Forests

Do you consider the Bank to be a global leader in forest-related matters?

Yes	27.5% (11)
No	60% (24)
Blank	12.5% (5)

Should the Bank be a global leader in forest-related matters?

Yes	45% (18)
No	25% (10)
Depends on issues	2.5% (1)
Blank	17.5% (7)

Based on your implementation experience, please indicate if you agree or disagree that the Bank is well positioned to provide leadership on the following global strategic issues in a policy and operational context.

	SA	A	D	SD	NA
i. Forest sector	22.5% (9)	42.5% (17)	22.5% (9)	5% (2)	7.5% (3)
ii. Climate change	12.5% (5)	52.5% (21)	7.5% (7)	2.5% (1)	15% (6)
iii. Biodiversity conservation	20% (8)	52.5% (21)	15% (6)	(0)	12.5% (5)
iv. Desertification	5% (2)	42.5% (17)	20% (8)	5% (2)	27.5% (11)
v. Natural resource management	20% (8)	57.5% (23)	7.5% (3)	2.5% (1)	12.5% (5)
vi. Carbon sequestration and the Clean Development Mechanism	20% (8)	35% (14)	20% (8)	2.5% (1)	22.5% (9)

SA: Strongly Agree; A: Agree; D: Disagree; SD: Strongly Disagree; NA: Not Available.

VIII. Other Issues

In general, do you think that the Bank's safeguard policies help or hinder forestry projects?

Help	40% (16)
Hinder	22.5% (9)
Both	7.5% (3)
Blank	30% (12)

Based on your experience, how do you rate the compliance with forest-related safeguard policies in the Bank's operations that impact on forest?

Highly satisfactory	2.5% (1)
Satisfactory	50% (20)
Unsatisfactory	12.5% (5)
Highly unsatisfactory	0
Blank	35% (14)

Do you think forest-related issues are sufficiently integrated in the Bank's agriculture sector ESW and strategy?

Yes	22.5% (9)
No	57.5% (23)
Blank	17.5% (7)
Depends on country	2.5% (1)

Do you think there is sufficient integration of forest sector issues in the Bank's work on poverty reduction?

Yes	7.5% (3)
No	75% (30)
Blank	17.5% (7)

Do you believe that the GEF's role in the forest sector should...

Increase	57.5% (23)
Stay the same	25% (10)
Decrease	2.5% (1)
Blank	15% (6)

Do you believe that the IFC's role in the forest sector should...

Increase	57.5% (23)
Stay the same	25% (10)
Decrease	2.5% (1)
Blank	15% (6)

THE OED SURVEY OF THE CEO FORUM

Sample size:	31
Responses:	16
Response rate:	52%
Date Sent:	April 20, 1999
Received:	May 3, 1999-October 5, 1999

Were you familiar with the World Bank's forest strategy before the CEO Forum was formed?

Yes	69%
No	31%

- 42% of the private sector respondents said "no."
- 100% of the NGO respondents said "yes."

How would you describe your current familiarity with the 1991 strategy?

Very familiar	25%
Somewhat familiar	75%
Not familiar	0%

- 75% of the NGOs said "somewhat familiar."

Do you believe the conservation focus of the forest strategy should continue?

Yes	69%
No	31%

- 100% of the NGOs said that the conservation focus should continue.
- 58% of the private sector respondents said the focus should continue.

Private Sector Comments:

- More broad-based approach to promoting sustainable forest management and sector reforms.
- The 1991 strategy did not sufficiently include the economic aspects of sustainable development. Emphasizing conservation as opposed to sustainable forestry will exacerbate the problem and continue to cause problems for the Bank with respect to credibility.
- Sustainable logging on a commercial scale is essential in development and good conservation and should not be ignored.
- The strategy should be made clear by dividing the forests into three parts:
  Conservation Forests – The Bank must help the states to define the areas and to protect them with a long-term view.
  Production Forests – The Bank must finance the research and help the operators in sustainable management.
  Conversion Forests – In agriculture, the Bank must finance research and aid to reduce shifting cultivation. Plantations are very important, but up to now it seems unclear if plantations should be cultivated in tropical moist forests or only in secondary forests or savannas.

Should the Bank's focus be directed toward the conservation of tropical moist forests?

Yes	50%
No	44%
No response	6%

- However, the majority (75% NGOs and 55% of the private sector) noted that all forest types should be covered.

Private Sector Comments:

- While tropical moist forests are essentially important, the conservation of forests worldwide is necessary. The issue should be addressed globally.

NGO Comments:

- Tropical moist forests may be the most critical, but is isn't useful to suggest that conserving one kind of forest addresses the need to conserve other kinds. Although most tropical moist forests are in poor countries, which is the Bank's main target, other forest types exist which also need conserving in Bank client countries.
- It would be useful to address conservation needs in other threatened forest ecosystems, and the Bank should consider broadening its strategy to cover all forest types.
- The strength of the Bank's commitment to direct conservation efforts should increase relative to related activities.

What are your views of the commercial logging ban in Operational Policy 4.36?

- The OP was criticized both for making too many exceptions to the policy and for not making enough in light of the assessed need for Bank involvement promoting sustainable forest management.
- Half of the private sector responses specifically advised the Bank to retain a presence in tropical moist forests. Yet this presence should be encouraged only if a high standard of sustainable forest management can be achieved. Also, to meet local and national stakeholders' needs, clearer definitions are needed to determine exactly what "as much forest cover as possible" means.

Private Sector Comments:

- The policy handicaps the Bank in its efforts. A lot of emerging and developing countries need the cash that comes from developing resources whereas they have very little other sources. The Bank had no clout to moderate conditions under which these areas are logged. The Bank can help countries move toward sustainable forest management if it is involved. This is a limiting factor in the Bank.

- We think every country has the right and obligation to use its resources in a reasonable manner to provide for the well-being of its population. That obligation includes the preservation, conservation, and utilization of the forest resources under an effective national forest policy that is itself part of the country's land resource allocation and development programs. The Bank's policy of banning participation in commercial logging activities merely shifts funding to other, often less responsible, sources. It removes the Bank and its direct influence from areas that most need its assistance to achieve complete and sustainable development programs. It is inconceivable to actually expect that all of a poor country's primary forests can or will be preserved. However, the most important aspect of maintaining appropriate forest ecosystems is to provide ecological, social, and economic alternatives to deforestation that address underlying causes of the phenomenon.
- Most of the tropical countries have land-use-planning in the sense that parts of the forests are completely protected (national parks) and other parts are earmarked for production. Part of these production forests are primary tropical moist forests. Especially in these forests where timber harvesting takes place in connection with or without the timber industry special attention.

NGO Comments:

- The Bank should seek the governments' commitments to the conservation of biodiversity by creating a functioning protected-area network before seeking a commitment to manage timber supply sustainably. The protection of biological diversity must be in itself an objective. Attempts to improve the planning, monitoring, and field growth of even secondary forests should proceed, but should not be the primary means of protecting biodiversity.
- Exceptions to the policy should only be allowed 1) for operations which have made a clear commitment to seek Forest Stewardship Council (FSC)-based certification and where there is a reasonable chance of this being achieved and 2) for community-based operations operating on a small scale.
- Financing commercial logging that is consistent with maintenance of high levels of ecological integrity may be good policy unless the blanket prohibition is needed because the bank is unable to exercise good judgment or penalize bad management and policy.

Do you see different roles for the small-, medium-, and large-scale timber producing countries?

Yes	81%
No	0
Don't know	19%

Private Sector Comments:

- Different sale opportunities/product line opportunities in different countries. Also, optimum resource allocation.
- Programs must be differentiated to take care of all these categories. Out-grower programs are important.
- Small companies manufacture specialty items better than large. They can handle smaller scales. In terms of tree planting, large companies tend to have economies of scale which tend to give them advantage. Where you can generate scale economies, there is some advantage to size.
- Small timber producers usually are nationals of the concerned tropical country. Their main role cannot be industrialization and downstream processing due to lack of funds and expertise. A small timber producer can supply his production to the local industry. A small timber producer cannot establish a management plan. He will not be able to manage his forest sustainably in the sense of a 30-40 year rotation. Small producers cannot give a guarantee for long-term employment. Due to lack of education, they are usually unable to train employees. The population does not find a permanent "home" and has to look for a new place (including shifting cultivation) when the suitable trees in a small concession have been harvested. Small producers: especially when they have bought the rights of using the forest on auction tend to "cut out and get out" method, which is exactly contrary to what is necessary.
- Large-scale companies have good possibilities for socially and environmentally more demanding and long-lasting development projects.

- Training can be done at the appropriate scale for large loggers using heavy equipment and small family operations that still need to apply good planning and management.
- The often more visible role of the large manufacturing company is in fact only part of the forest-based enterprise. Especially in solid wood products, there is an absolute need for small to medium- size producers who can adapt to local scale and circumstances, and these often are the most entrepreneurial business units. In a world in which the tendency is toward outsourcing of many services and parts of the manufacturing process, it is becoming commonplace for third party businesses to provide the road construction, harvesting and transportation and many of the silvicultural services related to forest business. A high level of quality practices and programs are necessary to provide benefits to all intended groups, to instill a feeling of responsibility for the forest in all who benefit from it, to account for concerns about equity, and to minimize distortions caused by those few who, inevitably, don't comply with the rules. These are all necessary components of building credibility for the projects.
- We are having industry provide materials for logging operations being done by subcontractors (smaller companies). The timber industry is a global market and on a global scale; one must discuss export. There are not enough local markets to sustain this (export) industry. First-grade export quality goes toward international export. One needs to understand regional sources and politics on a global level. Smaller companies can concentrate on logging operations under contracts of bigger companies or if they are industrial then they need good relations with three or four steady clients.

NGO Comments:

- Although roles vary from place to place, in general small- and medium-scale producers focus on supplying local markets with products involving little capital investment, such as rough-sawn boards. Large-scale producers generally focus on export markets and more capital-intensive production (e.g., plywood, veneer, furniture, etc.)
- Large-scale producers are able to engage in landscape-based planning and management. Smaller-scale producers need to group together to do this. Also, there are various other economies of scale which give commercial advantages to large-scale producers. Small-scale producers often benefit from a wide range of other goods/services.

From the perspective of your organization, do you see a benefit in pursuing certification?

Yes	56%
No	31%
Depends	6%
Don't know	6%

Is certification an effective tool for assessing specific standards of forest management?

Yes – given certain criteria	63%
No	13%
Yes and no, depends	25%

Is certification feasible given the current knowledge of forest management?

Yes	44%
No	13%
Yes and no, sometimes	31%
No response	6%

Is certification likely to be more successful in temperate than tropical forests?

Yes	63%
No	32%
Don't know	6%

Is certification likely to promote "sustainable forest management?" Please clarify your perception of sustainable forest management.

Yes	38%
No	31%
Depends on definition of sustainable forest management	19%
Too early to tell	6%
No answer	6%

– *Most members of the CEO Forum noted some benefits from certification but had a number of concerns about the costs and benefits and the willingness of end-users to pay significantly more for certified products, about certification becoming a barrier to trade, about problems of third-party monitoring, and about establishing a general agreement on guidelines.*

Private Sector Comments:

- As a tool, certification can serve as a motivator of excellence by identifying and recognizing those who excel in meeting standards of excellence. . . . In our experience, forestry certification schemes have not been the principal driver of improved forestry practice. It is effective in providing motivation and as a model, but it is too complicated a concept to reach the majority of forest owners in societies where the majority of forestland is owned by private individuals. Existing schemes are similar in terms of actual requirements; they differ in ideological agendas. . . In countries where there is little statutory or regulatory control and less general agreement about acceptable management practices, forest certification may be able to identify those forest enterprises that are properly meeting their responsibility for managing the forest resource. There is controversy about certification in the temperate regions (evidence that the process is not yet mature enough). Many forests that have been certified in many regions come nowhere close to our corporate standards.

- Certification does not touch on the real factors of deforestation in the tropics. Insensitive markets represent more than 95 percent of tropical wood production. Certification doesn't concern governments and cultivators, but only exporters to a few countries. Good management by all stakeholders—foresters, farmers, hunters—under the control of a strong administration has a greater effect on tropical forests than the good management of a few European companies who are already working in certifiable conditions.
- The FSC's criteria and indicators for certifying tropical natural forests are unrealistic and discriminating to tropical countries. There is not a single hectare of certified natural forests in West and Central Africa for the time being. If there are realistic criteria and indicators which leave the tropical country and its governments the sovereignty to accredit certifiers and use their own criteria and indicators (ATO - African Timber Association, CIFOR - Center for International Forestry Research), certification is feasible. The level of a sort of minimum criteria should be comparatively easy to reach by concession holders and timber industry. Criteria and indicators should allow a step-by-step process and set realistic targets which can be reached in a certain period of time. The monopoly of FSC and unrealistic criteria/indicators are a handicap in the certification process of tropical countries.
- The controversy about forest certification is primarily political in that it involves issues of ideology, control, financial advantage and starkly differing perceptions of what problems it is intended to solve and whether it can solve them. On the technical level, verification of forest practices to a particular standard is eminently feasible, if the standard is generally agreed upon. In the developed countries, where there is a well-developed framework of forest technology, laws and regulations, land use history, etc., it is entirely possible to define "generally accepted forest practices" that serve as a basis for certification. Much of this is potentially useful for forests in temperate and boreal developing countries. Some of it is equally useful in the plantation forests of the Southern Hemisphere. However, tropical rainforest forestry is in its infancy, it is largely confined to extraction of furniture-grade hardwoods, there is little silviculture and forest management as such, and there is little basis to establish agreed standards that have any practical technical basis.
- Certification can only influence those forests that produce products into a market that requires it. It will have no impact where the reasons for deforestation or degradation of an area are related to population expansion and agricultural or fuel needs. It is easy to certify temperate forests in countries with more than 200 years of forest research and sustainably managed forests. Forests in Northern Europe, for instance, have been managed sustainably for two centuries. Forestry in temperate zones (meaning rich industrial countries) cannot be compared with forestry in tropical zones (poor developing countries).

NGO Comments:

- (Not in tropical forests)—Sustainable forest management is presumably targeted (at a minimum) at sustained timber yields.
- Sustainable forest management is a long-term goal with ecological social and eco-dimensions. Its precise formulation will depend on what forest goods and services society wishes to "sustain." Certification can be a tool to promote this goal under appropriate circumstances.
- Sustainable forest management = ecologically compatible forestry—pursuit of management that protects biodiversity while permitting economic development. This could involve well-managed timber production and/or activities such as ecotourism or others.

Do you believe carbon markets can help in forest conservation?

Yes	75%
No	13%
Depends	13%

– *The NGO community unanimously believes that carbon markets can help in forest conservation.*

Private Sector Comments:

- The Kyoto Protocol to the United Nations Framework Convention on Climate Change has so much uncertainty associated with it that the form a carbon market can take is strictly speculative. It is difficult to see how a carbon sequestrat ion reserve of existing forest will provide the benefits needed to sustain a population that is using the area to meet their basic needs. This is especially true if the expectation is for "permanent" storage.
- The current concept of a base date and the limitations of the "Kyoto Forest" provide such significant disadvantages to the developed countries that it is difficult to see how they can ratify the treaty as currently conceived. At a recent meeting jointly sponsored by the Subsidiary Body for Scientific and Technological Advice (SBSTA) and the Intergovernmental Panel on Climate Change (IPCC), it was evident that several countries are just beginning to analyze the impacts of the current understanding on their own economies. However, regardless of the governmental context or political mechanism, forest plantations are an ideal way to increase the forest area for greater carbon sequestration. A study in Brazil, for example, indicated that significant reforestation with tree plantations of areas earlier converted to pasture and low-grade farmland would make a significant contribution to carbon sequestration while providing a long-term base for industrial wood. Similar opportunities exist through the Southern Hemisphere. A massive reforestation effort in India, parts of China, in Sub-Saharan Africa and Madagascar, would have enormous implications for carbon sequestration as well as for soil stability, local fuel and construction, and social improvement.
- It depends on whether funds created by carbon markets can be invested in tools of improved forest management which neither developing countries nor the private sector can finance.

Do you believe the climate change negotiations will result in clear guidelines for formulation of carbon trading markets?

Yes	19%
No	25%
Don't know/unclear so far/not anytime soon	50%
No response	6%

Should the World Bank play a role in the development of carbon markets?

Yes	81%
No	13%
Not sure	6%

In what ways can the World Bank and the private sector work together. . .

. . . to meet the demand for forest products?

- The World Bank can support educational efforts, information exchange, designed to spread sustainable forest management practices more broadly on a national and international basis. Also, research on trends in forest product markets.
- For all objectives, develop programs to demonstrate sustainable forest management/low impact logging. Also, develop forest plantations in "logged-over" areas.
- Lending of good projects in non-OECD countries.
- IFC could finance projects. The World Bank could improve the inventory climate and can play a positive role to motivate companies to follow sustainable forest management methods.
- By helping to establish the right environment for private sector investment in forest plantations (e.g., by providing concessional loans or by restricting non-plantation wood supplies by placing old-growth forests off-limits in protected areas).

. . . to help conserve primary forests?

- Education is critical here. These are public lands and the support of the public and the local community is vital to their preservation. Efforts at building coalitions that are broad in scope and purpose should be encouraged.
- Countries should set aside conservation areas; the rest of a country's area should be used economically without destroying it. Identify and protect private protected areas and collaboratively encourage governments to protect forests on public lands. Increase funding for protected area creation and management.

. . . to maintain biodiversity?

- Fund studies that promote sustainable forest management with the focus of maximizing biodiversity and make sure this information is widely available to industry so the principles can be broadly adopted.
- By understanding that sustainable management of primary forests is also possible, in connection with full conservation.
- Increase funding for protected area creation and management.

. . . to sequester carbon?

- Bring the players together and define the terms of this market.
- The World Bank could possibly help to develop carbon trading rules.
- Promote timber products and housing schemes.
- Fund studies to gain a complete and accurate understanding of carbon sequestration and develop incentives to accomplish this.

. . . to achieve other objectives?

- Work with governments to assure progressive, but practical, and enforceable policies that encourage improved forest management.
- Drive governments to good governance through cooperation with private industry. Where IFC is financing industry, the World Bank has a tool to promote discussion between governments, industry and the Bank. It seems as if the World Bank is ignoring the timber sector.
- Identify ways to develop and maintain standards of sustainable forest management through independent certification and other ways of verification.
- The World Bank (or IFC) could finance the improvement of state agencies, but also finance and work together with professional organizations that have pedagogic functions, like IFIA. The development of better methods in reduced-impact logging; communications and awareness of the private sector to improve these methods and pilot projects such as: forest management plans, environmental impact studies, low-impact logging, local processing with higher impact value, taxation and forest policy, and role of the forest industry in rural development.
- Sector reforms/investment climate, institutional development, HRD, financing, etc.

What contributions, in your opinion, has the CEO Forum made in the areas mentioned in this questionnaire?

- *The majority (60%) note a positive contribution.*
- *The rest believe that the Forum's contributions are unclear (27%) or limited (13%).*

Private Sector Comments:

- Essential discussion and information exchange forum, but a more active role is needed.
- There is some indication of convergence between the views of the NGOs and the CEOs, which is very positive. We have to go further by replacing the Forum with working groups that combine NGO/CEO/Private Sector/Bank, and the like.
- The CEO Forum is the first move of the World Bank toward better understanding between the Bank and the private sector as "motor" for development in general and in special for forest management, industrialization, education, creation of jobs, fight against poverty, etc.
- It has made a small group of industry CEOs more aware and better informed about the objectives of the World Bank and the "mindset" of its leadership. It has provided the foundation for a small group of companies and NGOs to initiate a limited (and somewhat superficial to date) dialogue on certain aspects of sustainable forestry. However, those actions have helped to foster a framework in which a broader dialogue can continue among the industrial, private forest owner and NGO communities, and that is a worthwhile accomplishment.
- Relatively little, so far.
- None yet, it is a talking shop.

NGO Comments:

- Unclear as yet, as working groups are still in progress.
- Brought together different sectors for exchange of views/made some progress through the working groups.
- Brought key parties together for important discussion and awareness raising of each others' perspectives.

Do you believe that the IFC's role in the forest sector should . . .

Increase	56%
Decrease	0
Stay the same	6%
Don't know	25%
Depends on focus	13%

- *The NGO community felt that IFC's role should "stay the same" or should be increased in situations where it uses good ecological judgment.*

Private Sector Comments:

- IFC's role should increase in non-OECD countries and in some newly established members of OECD.
- IFC knows the forest sector and understands investment climate in emerging markets.
- The IFC should support commercial plantations.
- It is my impression and experience that the IFC has more or less pulled out of investments in the forest sector. IFC and also MIGA could play a better and more constructive role for developing the forest sector and timber industry in a sustainable way in the future.
- IFC could finance some special cooperation like: professional training to promote specialists in forest inventory, sustainable management, logging, sawmilling, etc. Or, the financing of equipment of national interest like main roads (e.g., Central Africa-Douala) and railways (Congo, RD Congo, Cameroon, etc.)

NGO Comments:

- The IFC should not be involved in forestry projects in the humid tropics, nor should they be lending for projects such as roads and pipelines having significant negative impacts on tropical forests. Rather, the IFC should increase lending for plantation development on degraded land and to conservation-friendly commercial development activities.

Do you believe that the GEF's role in the forest sector should . . .

Increase	44%
Decrease	0
Don't know	38%
Stay the same	6%
Depends on focus	6%
No response	6%

- *50% of the private sector respondents were not able to answer the question, usually for lack of knowledge about the organization.*
- *66% of the NGO community supported an increase in GEF activity.*
- *The other respondents noted that while some areas of GEF involvement could be heightened, its work should be "confined to projects whose conservation and environmental benefits are beyond question, such as funding protected areas through a trust fund."*

Private Sector Comments:

- GEF should increase its financial involvement in natural resource management projects, changes in operational practices, and (especially) financing agriculture and animal breeding, which have to be integrated into forest management, industrial development with fast-growing population in connected villages. This issue has been completely neglected up to now, at least in West and Central Africa. It is absolutely necessary to combine logging operations, timber industry (result = fast growth of villages) with agriculture and animal breeding projects.
- GEF should become more involved in programs to demonstrate and teach sustainable forest management/low-impact logging.
- GEF should increase activity in non-OECD countries or newly established members of OECD.

Based on your knowledge of the Bank, do you believe the Bank is well positioned to address global strategic issues such as the role of forests in climate change, desertification, biodiversity conservation, and resource management?

- *Overall, 73% of all respondents felt that the World Bank is well positioned to address the issues of climate change, biodiversity conservation, and resource management.*
- *60% of all respondents felt that the Bank is well positioned to address desertification.*
- *The NGO community was critical of the World Bank's role in addressing these global issues.*

NGO Comments:

- The Bank should primarily play a "facilitator's" role in issues related to forests and climate change and for biodiversity concentrate more on providing grants to private sector partners and in helping to fund implementing agencies in developing countries.
- The Bank has difficulty with multi-country programs and is therefore not well positioned to address any of these issues.
- While the Bank could address biodiversity conservation and resource management, it is not well equipped to address climate change and desertification.

THE OED SURVEY OF THE WB/WWF ALLIANCE

Sample Size:	93
Responses:	34
Response Rate:	37%
Date Sent:	May 3, 1999
Received:	May - September 1999

In what capacity and in what countries are you involved in the WWF-World Bank alliance? On average, how many hours a month do you dedicate to the Alliance?

	Bank		WWF	
	Title	**Hours per month**	**Title**	**Hours per month**
AFR	Task manager/natural resource management specialist	2	Regional point person	40
	Task team leader/forests LS	16	Program officer	15
	Bank regional manager	10	Country team member	8–12
	Senior environment economist	2		
EAP	Region coordinator	5	Regional point person	50
	Resident mission in China	less than 5		
	Task manager	minimal		
ECA	Biodiversity specialist	2	Director	1
			Forest officer (ECA Program)	20
			Forest officer (Mediterranean Program)	10
			Forest officer (Russia)	8–12
LCR	Biodiversity specialist	0	Forest coordinator	over 100
	Senior economist	8	Conservation director	4
	Senior forest specialist	5	Forest officer	30
	Task manager/economist	0.5	Technical director	6
	Regional manager	0	Country team member	3
	Senior natural resource management specialist			
	Sector leader	2		
SAR	Task manager	less than 5	Program officer	5–10
	Task manager	35–40		
	Senior environment specialist	0		
	Forestry specialist	2		
	Senior anthropologist in social development department	0		

From your perspective, what are the reasons for the partnership between the World Bank and the WWF?

Bank		WWF	
Response	Frequency	Response	Frequency
Partnership exists to combine the strength and expertise of both organizations (to utilize the comparative advantage of both institutions).	15	WWF ensures that environmental and conservation issues are accorded the desired importance by key actors, players, and decisionmakers.	6
An association with an environmental NGO, given the controversial nature of forest management, is good public relations for the Bank.	5	An opportunity for WWF to influence WB agenda and programs related to protected areas and the forest sector.	5
Partnership affords WWF the opportunity to attempt to influence the Bank's forestry portfolio in that it realizes that the Bank's management does not give it enough attention.	3	Additional funding for WWF.	1
WWF will benefit from increased funding possibilities and that the institution will gain more clout through association.	2	An exploratory attempt to conciliate two frequently opposed view: development and conservation.	1

What are the benefits of the Alliance?

Bank		WWF	
Responses	Frequency	Responses	Frequency
Synergy, shared goals, and blending of cultures	3	Increase in funding for conservation activities	4
No benefits	3	Increased attention to conservation and sustainable forestry issues and the promotion of constructive uses of resources	4
Provide most funds, money from the center for forestry activities, an increase in resources	3	Synergy and the establishment of a common framework	3
Increase representation of key stakeholders in Bank operations	2	Allows WWF more direct access to World Bank staff	2
Promote forest conservation and best practices in forest management	2	The chance for the World Bank to have more influence over the private sector and to have more non-typical World Bank activities	2
Political clout	1	The Alliance targets now to be "attacked" from both directions	1
Raise awareness on social, economic and environmental benefits of management and conservation forests	1	No benefits	1
Provide a medium for governments to assume their responsibility of forest management	1	In practice, hard to say at this point, except that governments may be taking more notice of forest conservation issues if the Bank's name is associated	1
Propaganda	1		
More visibility to forestry with Bank managers	1		
No answer	2		

What are its drawbacks?

Bank		WWF	
Responses	**Frequency**	**Responses**	**Frequency**
The Alliance is top down.	5	The level of motivation and commitment from the WB staff to make the Alliance work is much lower than WWF. Lack of staff time.	2
The cultures and objectives of WWF and WB are not perfectly matched.	5	WWF is giving credibility to the Bank with little guarantee of effective action on the part of the Bank. For WWF, close association with the WB, if not well explained, could pose some problems in terms of image, particularly due to past negative environmental impacts of WB projects worldwide.	2
No drawbacks.	4	Overloading an already stretched system.	1
Modest funds are too small to be cost-effective for Bank staff.	2	Lack of "buy-in" by other partners.	1
False expectations.	2	Turf Matters—coordination difficulties.	1
• Differences in financing mechanisms.	1	Institutional cultures, roles, and physical distribution of staff are very different.	1
• Differences in sectoral interests.	1		
• New alliances are next to useless unless they come with new funding and that was the problem with the WB/WWF Alliance.	1		
• Not enough attention given to interests of indigenous people.	1		
• No "buy in" amongst clients or within the Bank.	1		
Seen as Bank giving in to FSC agenda and WWF agenda.	1		
		Insufficient emphasis on addressing "paper parks" relevant to the target of creating new protected areas.	1

Do you believe the goal of establishing 50 million hectares (125 million acres) of new forest protected areas is a realistic target?

Bank		WWF	
Responses	**Frequency**	**Responses**	**Frequency**
Yes	8 (40%)	Yes	12 (86%)
No	12 (60%)	No	0
Don't know	0	Don't know	1 (7%)
No answer	0	No answer	1 (7%)

Is progress toward the realization of this target likely to be made in temperate forests?

Bank		WWF	
Responses	**Frequency**	**Responses**	**Frequency**
Highly likely	0	Highly likely	3 (21%)
Likely	14 (70%)	Likely	7 (50%)
Unlikely	1 (5%)	Unlikely	1 (7%)
Highly unlikely	0	Highly unlikely	0
No answer	5 (25%)	No answer	3 (21%)

Is progress toward the realization of this target likely to be made in tropical forests?

Bank		WWF	
Responses	Frequency	Responses	Frequency
Highly likely	0	Highly likely	6 (43%)
Likely	7 (35%)	Likely	3 (21%)
Unlikely	4 (20%)	Unlikely	0
Highly unlikely	4 (20%)	Highly unlikely	1 (7%)
No answer	5 (25%)	No answer	4 (29%)

Do you believe the goal of bringing an additional 200 million hectares (500 million acres) of the world's forests under independent certification by the year 2005 is a realistic target?

Bank		WWF	
Responses	Frequency	Responses	Frequency
Yes	6 (30%)	Yes	9 (64%)
No	11 (55%)	No	4 (29%)
No answer	3 (15%)	No answer	1 (7%)

Is progress toward the realization of this target likely to be made in temperate forests?

Bank		WWF	
Responses	Frequency	Responses	Frequency
Highly likely	1 (5%)	Highly likely	8 (57%)
Likely	12 (60%)	Likely	3 (21%)
Unlikely	1 (5%)	Unlikely	0
Highly unlikely	0	Highly unlikely	0
No answer	6 (30%)	No answer	3 (21%)

Is progress toward the realization of this target likely to be made in tropical forests?

Bank		WWF	
Responses	Frequency	Responses	Frequency
Highly likely	0	Highly likely	3 (21%)
Likely	7 (35%)	Likely	7 (50%)
Unlikely	4 (20%)	Unlikely	2 (14%)
Highly unlikely	4 (20%)	Highly unlikely	0
No answer	5 (25%)	No answer	2 (14%)

From your perspective, how do your next-in-line managers view the Alliance?

Bank		WWF	
Responses	Frequency	Responses	Frequency
Highly desirable	1 (5%)	Highly desirable	5 (36%)
Desirable	2 (10%)	Desirable	3 (21%)
Somewhat desirable	5 (25%)	Somewhat desirable	5 (36%)
Undesirable	5 (25%)	Undesirable	0
No answer	2 (10%)	No answer	1 (7%)
Managers don't have an opinion/ uninterested	5 (25%)		

Who, in your view, are currently the stakeholders of the Alliance within the Bank?

Bank		WWF	
Responses	Frequency	Responses	Frequency
Env anchor/staff	8	Env and Social Policy Division	4
RDV Family (Forestry)	6	Wolfensohn	3
Wolfensohn	4	Forestry Division	3
Ken Newcombe	2	Country management units	2
Select TTLs	1	GEF coordinating staff	1
A few naïve TMs	1	Regional point people	1
External affairs	1	LCR staff	1
None	1		

Which additional key stakeholders within the Bank, in your opinion, need to be involved in the Alliance to ensure its successful implementation?

Bank		WWF	
Responses	Frequency	Responses	Frequency
Country directors	5	Country directors/desks	5
ENV and RDV sector managers	2	Regional offices/desks	4
Vice presidents	2	Bank operational staff	3
All TMs working on GEF	2	Vice presidents	2
Middle-level managers	2	Task managers	2
Environment department	2	Biodiversity Thematic Group	1
The Regions	1	Environment	1
Regional management	1	Regional directors	1
Social Development Family	1	Resident representatives	1
Operational VPs	1	RDV	1
Natural resource task team	1		
External relations managers	1		
Technical managers	1		
Private sector groups	1		
Fundraisers	1		
In the EAP Region: the RVP and the sector manager responsible for "regional initiative	1		
Biodiversity Park Project and Forestry Project	1		
IFC	1		

Who, in your view, are the current stakeholders outside the Bank and WWF?

Bank	WWF
Responses	Responses
International donor community	Ministry of Environment
DFID	Ministry of Agriculture
Danida	Governments
Netherlands	Academic and research institute
UNDP	International Development Agencies
FAO	International Cons. and Dev., NGOs
IUCN	International organizations
UNEP	FSC
UNIDO	Forestry Organizations: Soc. of Tropical Foresters
Governments	None
Some Forestry Departments	Industry/ private sector
None	Local NGOs
Industry	National NGOs
Social groups	Local communities
Environmental NGOs	
Africa: Cameroon and Madagascar	
Consultants involved with certification	
Indigenous communities	
Tribal communities	
Private sector	

Which additional key stakeholders outside the Bank and WWF need to be involved in the Alliance to ensure its successful implementation?

Bank		WWF	
Responses	Frequency	Responses	Frequency
Local NGO networks, national (client) NGOs	7	National governments	5
Government agencies responsible for forest management	6	Local communities	4
Indigenous groups	2	Private Sector dealing with forest resources	4
Local communities	2	Indigenous groups	3
Academia	2	International NGOs	3
Other NGOs	2	International organizations/ AID agencies	2
International NGOs (WRI, IUCN)	1	Women's organizations	1
Ministers of Planning	1	Research institutions/forestry schools	1
Ministers of Finance	1	National forest administrator	1
National governments	1	Ministers of Finance	1
Ted Turner	1	Small business	1
Politicians	1	Local governments	1
Bureaucrats	1	The EU MEDA/ SMAP	1
		FUNDECOR, CODEFORSA	1

Do you think that there is currently adequate involvement of key in-country Stakeholders in the Alliance?

Bank		WWF	
Responses	Frequency	Responses	Frequency
Yes	0	Yes	1 (7%)
No	18 (90%)	No	12 (86%)
No answer	2 (10%)	No answer	1 (7%)

In what ways is the Alliance contributing to the Bank's institutional mission: "poverty alleviation and sustainable development"?

Bank Responses	Frequency	WWF Responses	Frequency
Too early in the game to tell.	4	Alliance targets are conducive to both mission goals.	2
People are mainly lost in Alliance debates.	4	The certification target is relevant to sustainable development.	2
Only marginally contributing to the Bank's mission.	2	Biodiversity conservation is a long-term requirement for sustainable development	2
It is unclear if it contributing to it.	2	Certified forests should contribute to more competitive forest products industries, improve labor practices, and address human rights issues.	1
Alliance is not contributing to the Bank's overall mission.	1	No tangible ways yet.	1
No appreciable contribution of the Alliance in South Asia.	1	Where the Alliance helps realize sustainable forest use, local people should benefit and their standards of living should be improved. The benefits are less direct with regard to forest protected areas, but effects such as watershed protection and biodiversity conservation are significant.	1
Developing more inclusive partnerships with those concerned with sustainable management of natural resources.	1	Protected areas are a key to biodiversity conservation and a key component of sustainable development.	1
Certification is contributing to the WB mission because it should ensure sustainability and higher income for the poor.	1	Sustainable natural resource management is a prerequisite for creating the enabling environment for the Bank mission.	1
To the extent that certification is broadly defined, can increase transparency and accountability on the part of the resource managers we are contributing to both of these goals.	1	In the Mediterranean region, the fight against desertification and the conservation of water resources are both related to forest conservation and are fundamental for poverty alleviation.	1

In what ways is the Alliance contributing to the WWF's institutional mission?

Bank Responses	WWF Responses
Political clout	The alliance will help WWF deliver its forest conservation strategy.
Fundraising	The Alliance targets are linked with WWF Forest for Life campaign targets.
A way to add value to WWF's Forest for Life Campaign and other global initiatives	Additionally, alignment with the Bank is allowing us to be partners in the development of much larger projects, whose success could significantly improve the prospects for biodiversity conservation and sustainable resource management directly, as well as modifying national policies.
In some countries, WWF is being taken off-task when asked to address poverty alleviation and sustainable forestry instead of protected area management.	The Alliance has facilitated government commitment to creating new protected areas but has contributed very little in terms of actual implementation.
No answer or "don't know" (50%)	No answer (7%)

Given the diverse conditions among regions and countries, do you believe that sufficient conditions and incentives can be created to achieve a certification standard which will be compatible with globally applicable principles?

Bank		WWF	
Responses	Frequency	Responses	Frequency
Yes	8 (40%)	Yes	13 (93%)
No	7 (35%)	No	0
Not relevant	1 (5%)	Not relevant	1 (7%)
Undecided	4 (20%)		

Which type of timber producer is more likely to be more receptive to certification?

Bank		WWF	
Responses	Frequency	Responses	Frequency
Small	2 (10%)	Small	1 (7%)
Medium	3 (15%)	Medium	2 (14%)
Large	10 (50%)	Large	8 (57%)
No answer	4 (20%)	No answer	2 (14%)
Small and large	1 (5%)	Small and large	1 (7%)

Can certification be applied as an instrument of Bank forest strategy (that is, as a term of conditionality and as a tool for risk management)?

Bank		WWF	
Responses	Frequency	Responses	Frequency
Yes	10 (50%)	Yes	12 (86%)
No	9 (45%)	No	0
Maybe	1 (5%)	Maybe	0
No answer	0	No answer	2 (14%)

Is certification likely to be more successful in temperate than tropical moist forests?

Bank		WWF	
Responses	Frequency	Responses	Frequency
Yes	10 (50%)	Yes	9 (64%)
No	7 (35%)	No	1 (7%)
Depends (Depends entirely on markets and whether temperate wood markets are demanding such a system.)	1 (5%)	Depends	0
No answer	2 (10%)	No answer	4 (29%)

Is certification likely to promote "sustainable forest management"? Please briefly identify what you understand by "sustainable forest management."

Bank		WWF	
Responses	Frequency	Responses	Frequency
Yes	11 (55%)	Yes	9 (64%)
No	3 (15%)	No	1 (7%)
Maybe	1 (5%)	Too early to tell	3 (21%)
No answer	5 (25%)	No answer	1 (7%)

What three key constraints are you facing in the implementation of the Alliance?

Bank Responses	WWF Responses
Funding/high transaction costs for small WB/WWF portfolio.	Lack of clarity on who should be counterpart contact—lack of coordination, communication/ lack of information in the Bank/ lack of well-defined management structure.
Limited human resources/expertise to WWF and Bank.	Funding/inadequate resources/ staff/ technical support.
Lack of team building/absence of country-based activities and partnership/ lack of in-country demand and understanding.	Different modes of operation/types of projects/ lack of policy strategy/ lack of consensus on forestry policies.
Insufficient involvement of regional management and CD's.	Lack of Bank staff commitment/ lack of Bank acceptance of Alliance.
WWF's own priorities/ lack of WWF interest locally/ WWF national agency not agreeing with proposal.	Unclear procedural steps in how to become involved in current initiatives, or implement in the field, difficulties in establishing working relations in the field.
Lack of decisions on key technical and implementation issues/highly centralized decisionmaking.	Lack of commitment on both sides to targets/inadequate policies to support the two targets.
Time.	Government position/weak government forest service.
Lack of information.	Time.
Government foot-dragging.	Position of radical NGOs.
Lack of other stakeholders to assist with targets.	Largely unknown in-country (Bolivia).
Overcoming other NGO resistance.	
No consensus on "independent certification."	
No clear procedures on how to have a project considered an "Alliance project."	

What key lessons have you learned since the Alliance was started?

Bank Responses	WWF Responses
I must better prioritize my time and avoid unfunded mandates.	Where there are committed individuals on both sides, the Alliance has a greater chance of working to achieve mutually agreed goals. Where there is commitment of only one part, no.
Don't try to stretch money—money doesn't stretch.	The WB staff are too busy to participate in the Alliance everyday activities.
New initiatives require an initial investment of faith, time, and resources that the Bank is not always ready to make.	Communication channels within WWF cannot be relied on.
Managing expectations can be difficult as managing outputs and outcomes.	Clarity at conceptual and operational level is a prerequisite for developing a viable program; locate the decisionmaking points as close to the ground as possible.
The WWF has its own agenda, it is not interested in Bank's agenda except to influence it. "Partners" seem to be job-hunting in the Bank.	It is a lot more time-consuming if there is not more coordination among WWF-WB at high and regional levels.
None.	Funding is a prerequisite, but not by itself sufficient to achieve success.
Need to bring local forest dwellers and those that surround forests into the discussions as key participants.	Financial resources are a secondary issue for success of joint initiatives.
More education for the Alliance will help government agencies to better understand the Alliance.	There is a big problem of funding activities (the mechanism is still not working efficiently).
Bring other NGOs into an Alliance project. Need to enhance impact by creating partnerships.	Other stakeholders need to be brought into the Alliance at the planning stage."
WWF is not one partner but several (country offices, Washington and Gland).	"Fine words butter no parsnips." It was disappointing that President Wolfensohn did not attend the Yaoundé Summit, which would have added clout and credibility to the Alliance.
The Alliance is hampered by its construction around two partners who are largely external where forest conservation is concerned.	Certification may be a hard concept to explain in a developing country context. It may be easier to focus initially on the best forest management practices that lead, increasingly, to sustainable management.
Too top- down. Get an assistant or else never get involved in an initiative that came from top-down and lacks strong admin back- stopping. Bank management is not always ready for the implementation policies and initiatives decided upon by the president.	Governments of the African region have the final say. The commitment from government is crucial.
Focus on smaller number of (larger) initiatives.	The two resources CAN pull together their resources and strengths to achieve the targets.
	It will take time for the two organizations to learn to work together.

Do you have suggestions to modify, adapt or strengthen this initiative?

Bank	WWF
Responses	**Responses**
Streamline access to funds/commitment of adequate resources/ or advise clients that there is not any significant funding available, false expectations lead to dispointed clients and overworked task managers.	Other targets should be identified.
Decentralization with individual country focus.	We need to do a better sell of the Alliance at the regional level.
Need long-term planning.	Both targets need to be clarified, and, if possible, broken down to the regional levels.
Task managers in both organizations need to share a felt need for the Alliance.	A vision for the future is needed: what happens after the targets are met?
Hold more workshops to introduce the Alliance.	Communications and reporting lines should be more clearly defined, and should be observed by all levels of both organizations.
Focus on the goal of forest and other natural resource development.	Support and "buy-in" by other partners.
Regionalize 97 percent of contact between the parties and get rid of the big ambitious targets.	What does the Alliance mean for countries where only one of the organizations is active?
Give some thought to Miombo woodlands and other forest types in S. Africa region. It is vital to reinforce SADC forestry and biodiversity institutions and form wider partnerships (IUCN, SASUG, etc.).	A centralized team or pool of persons/experts that could travel to various countries trying to implement the Alliance and could facilitate/lobby for initial efforts/workshops regarding forest protected areas and certification/SFM would be useful.
Place greater responsibility on our client countries.	The Alliance materials should be translated in to the country languages.
Open it up to other national and international organizations.	Set aside specific grant funds to implement protected area targets.
We should not continue to spend any more effort until a review of value-added of such partnership to both WB and WWF country programs addressing sustainable forest management.	All actors need to internalize alliance goals and objectives if mainstreaming of alliance is to be achieved. It has to be fully integrated and accepted by the WB.
	More interest for cooperation in the Med area.

ANNEX H: STUDY STAFF AND CONSULTATION CONTACTS

Study Staff

Core Team
Uma Lele, Task Manager
Syed Arif Husain
Maisha Hyman
Lauren Kelly
Nalini Kumar
B. Essama Nssah
Aaron Zazueta

Additional Members
Madhur Gautam
Ridley Nelson

Consultants
Caroline Barnes
Madelyn Blair
Arnoldo Contreras
Kavita Gandhi
Karin Perkins
Saeed Rana

Authors of Supporting Studies

BRAZIL: *"Forests in the Balance: Challenges of Conservation with Development"*
Evaluation Country Case Study Series

Uma Lele, OED
Virgilio Viana, Professor of Forestry, Universidade de São Paulo
Adalberto Verrisimo, EMBRAPA, Amazon
Stephen Vosti, Visiting Scholar, Department of Agriculture and Resource Economics, University of California, Davis
Karin Perkins, Consultant, OED
Syed Arif Husain, Consultant, OED

CAMEROON: *Forest Sector Development in a Difficult Political Economy: An Evaluation of Cameroon's Forest Development and World Bank Assistance*

Boniface Essama Nssah, OED
Jim Gockowski, scientist (agricultural economist), International Institute of Tropical Agriculture, Cameroon

CHINA: *From Afforestation to Poverty Alleviation and Natural Forest Management*
Evaluation Country Case Study Series

Scott Rozelle, Associate Professor, Economics, University of California, Davis
Jikun Huang, Chinese Academy of Agricultural Sciences
Syed Arif Husain, Consultant, OED
Aaron Zazueta, Consultant, OED

COSTA RICA: *Forest Policy and the Evolution of Land Use*
Evaluation Country Case Study Series

Ronnie de Camino, President, Tropical Natural Resources, Inc. (RNT)
Olman Segura, Professor, Universidad Nacional
Luis Guillermo Arias
Isaac Perez, consultant with IDB

INDIA: *Alleviating Poverty Through Forest Development*
Evaluation Country Case Study Series

Nalini Kumar, OED
N. C. Saxena, Secretary to Government of India, Rural Development Department, New Delhi
Y. K. Alagh, Member of Parliament (Upper House), India
Kinsuk Mitra, Natural Resources Management Coordinator, Winrock, Inc.

INDONESIA: *The Challenges of World Bank Involvement in Forests*
Evaluation Country Case Study Series

Madhur Gautam, OED
Uma Lele, OED
Hariadi Kartodiharjo, Faculty of Forestry, Bogor Agricultural University, Institut Pertanian Bogor, Darmaga, Indonesia
Azis Khan, Researcher, Agency for Research Development, Indonesian Ministry of Forestry
Ir. Erwinsyah, Associate, Industrial Based Forestry Management, NRM Program, USAID
Saeed Rana, Consultant, OED

GEF Report: *Financing the Global Benefits of Forests: The Bank's GEF Portfolio and the 1991 Forest Strategy*

J. Gabriel Campbell, Consultant
Alejandra Martin, Consultant

IFC Report: *"OEG Review – Implementation of the 1991 Forest Strategy in IFC's Projects"*

Afolabi Ojumu, CEXOE
Rafael Dominguez, CEXOE
Cherian Samuel, CEXOE
Dominique Zwinkels, Consultant, CEXOE
John Gilliland, Consultant, CEXOE

Advisory Committee Members

Conor Boyd
President, Weyerhaeuser Forestland International

Angela Cropper
Chair, Editorial Committee, World Commission on Forests & Sustainable Development

Emmy Hafild
Director, WAHLI
Chair, Indonesian Working Forum (NGO/community organization)

Hans Gregersen
Chair, CGIAR Impact Assessment and Evaluation Group
Professor, College of Natural Resources, University of Minnesota

List of Nongovernmental Organizations (NGOs) Briefed

Africa Resources Trust
AGIR ICI, France
Bank Information Center
Biodiversity Action Network
Bionet
Campagna per al Reforma della Banca Modiale, Italy
Center for International Environmental Law
Center for Tropical Forest Science/Smithsonian Tropical Research Institute
Centre pour l'Environnement et le Developpement, Cameroon
Conservation International
Consumers Choice Council
Cousteau Society
Environmental Defense Fund
Evergreen Indonesia
Fern, Belgium
Forest Peoples Programme, UK
Forest Stewardship Council
Greenpeace International
Global Forest Policy Project
Indian Institute of Bio Social Research and Development
Indonesian Ecolabeling Institute
Indonesian Forum for Environment/Friends of the Earth Indonesia
Institute for Global Environment Strategies
International Institute for Energy Conservation
International Union for Conservation of Nature and Natural Resources
IUCN-Netherlands
IUCN-Washington
Japan Center for a Sustainable Environment and Society
Political Economy Research Center
Rainforest Action Network
Rainforest Foundation, UK
Ramkrishna Mission Lokashiksha
The Knowledge Initiative
Union of Concerned Scientists
W. Alton Jones Foundation
World Economy, Ecology and Development, Germany
World Rainforest Movement, Uruguay
World Resources Institute
World Wide Fund for Nature/World Wildlife Fund

LIST OF BILATERAL/MULTILATERAL INSTITUTIONS AND CONTACTS

Bilateral/multilateral institution	Contact
Center for International Forestry Research (CIFOR)	Jeff Sayer (j.sayer@cgiar.org) David Kaimowitz (d.kaimowitz@cgiar.org) Reider Persson (r.persson@cgiar.org) Ravi Prabhu (cgiar@worldbank.org)
Empresa Brasileiria de Pesquisa Agropecuária (EMBRAPA)	Francisco Reifschneider (sci@sede.embrapa.br)
Food and Agriculture Organization (FAO)	Michael Martin (Michael Martin@fao.org) Lennart Ljungmen (Lennart.Ljungmen@fao.org) Arnoldo Contreras (Arnoldo.Contreras@fao.org)
Deutsche Gesellschaft fur Technische Zusammenarbeit (GTZ)	Cornelis Baron van Tuyll van Serooskerken (Cornelis.Tuyll@gtz.De)
International Union of Forest Research Organizations (IUFOR)	Jeff Burley (jburley@plant-sciences.oxford.ac.uk)
International Center for Research in Agroforestry (ICRAF)	Pedro Sanchez (icraf@cgiar.org) Tom Tomich (t.tomich@cgnet.com) Erick C. M. Fernandes (icraf@cgiar.org)
Intergovernmental Forum on Forests (IFF)	Ilkka Ristimäki (ilkka.ristimaki@formin.mailnet.fi) Jagmohan Maini (maini@un.org)
International Food Policy Research Institute (IFPRI)	Steve Vosti (s.vosti@cgnet.com)
International Institute of Tropical Agriculture (IITA)	Lukas Brader (iita@cgiar.org)
Regional Unit for Technical Assistance (RUTA, Costa Rica)	James Smyle (jsmyle@ruta.org)
Swiss Agency for Development and Cooperation (SDC)	Theo Weiderkehr (t.weiderkehr@deza.admin.ch)

MAJOR MEETINGS/WORKSHOPS/REVIEWS

December 18, 1998	Entry Workshop – OED Review of the World Bank Group's 1991 Forest Strategy and Its Implementation (First Meeting of the Advisory Committee)
January 29, 1999	NGO Workshop
April 26–27, 1999	Second Meeting of the Advisory Committee for the OED Review of the World Bank Group's 1991 Forest Strategy and Its Implementation
June 29, 1999	One-Stop Review – Costa Rica Case Study Draft
July 27–28, 1999	Two-Day Forestry Retreat: • One-Stop Review meetings for Brazil, Cameroon, China, and India reports • Presentations on the Portfolio Reviews for ECA and LCR regions • Progress Report on Indonesia country study • Findings of the IFC and GEF reviews and the WWF questionnaires • MIGA contribution to the OED Review • Update on the OED Review outline
November 1–2, 1999	India Country Workshop – New Delhi, India
November 5, 1999	China Country Workshop – Beijing, China
November 18, 1999	Brazil Country Workshop – Brasilia, Brazil
November 22–23	Third Meeting of the Advisory Committee for the OED Review of the World Bank Group's 1991 Forest Strategy and Its Implementation
December 15, 1999	Briefing for Bank Staff
December 23, 1999	CODE Seminar • Briefing for Mr. James Wolfensohn
January 27–28, 2000	Forestry Review Workshop
April 25, 2000	Indonesia Country Workshop – Jakarta, Indonesia
February–May 2000	Regional Consultations – ESSD
June 2000	Report to CODE for Final Review

INDICATORS FOR ENVIRONMENTAL MANAGEMENT

Principle, criterion, indicator	Assessment of CIFOR criteria and indicators					
	Goal			Intervention point		
	Efficiency	Equity	Sustainability	Stand	Institution	Policy
Policy, Planning and Institutional Framework Are Conducive to Sustainable Forest Management						
There is sustained and adequate funding for the management of forests.	X	X	X		X	
Policy and planning are based on recent and accurate information.			X		X	
Effective instruments for intersectoral coordination on land use and land management exist.	X		X		X	
There is a permanent forest estate (PFE) adequately protected by law, which is the basis for sustainable management, including both protection and production forest.	X		X		X	X
There is a regional land use plan or PFE which reflects the different forested land uses, including attention to such matters as population, agricultural uses, conservation, environmental, economic and cultural values.	X	X	X		X	X
Yield and Quality of Forest Goods and Services Sustainable						
Management objectives clearly and precisely described, documented, and realistic.	X	X	X	X	X	
Objectives are clearly stated in terms of the major functions of the forest, with due respect to their spatial distribution.	X			X		
A comprehensive forest management plan is available.	X	X	X	X	X	
Maps of resources, management, ownership, and inventories available.	X			X		
Silvicultural systems prescribed and appropriate to forest type and produce grown.	X			X	X	
Yield regulation by area and/or volume prescribed.	X			X	X	
Harvesting systems and equipment are prescribed to match forest conditions in order to reduce impact.	X			X	X	
The management plan is effectively implemented.	X			X		
Pre-harvest inventory satisfactorily completed.	X			X		
Infrastructure is laid out prior to harvesting and in accordance with prescription.	X			X		
Reduced-impact felling specified and implemented.	X			X		
Skidding damage to trees and soil minimized.	X			X		
An effective monitoring and control system audits management's conformity with planning.	X		X	X	X	
Continuous forest inventory (CFI) plots established and measured regularly.	X			X	X	
Documentation and records of all forest management activities are kept in a form that makes it possible for monitoring to occur.	X		X	X		
Worked coupes are protected (e.g., from fire, encroachment, and premature reentry).	X		X	X	X	

INDICATORS FOR ENVIRONMENTAL MANAGEMENT (CONT'D)

	Assessment of CIFOR criteria and indicators					
	Goal			Intervention point		
Principle, criterion, indicator	Efficiency	Equity	Sustain-ability	Stand	Institution	Policy
Tree marking of seed stock and potential crop trees.	X			X		
Maintenance of ecosystem integrity						
Processes that support and maintain biodiversity of forest ecosystem are protected or enhanced.	X		X	X	X	
Endangered plant and animal species are protected.			X	X	X	
Interventions are highly specific, selective, and are confined to the barest minimum.	X			X		
Canopy opening is minimized.	X			X		
Enrichment planting, if carried out, should be based on indigenous, locally adapted species.	X			X		
The capacity of the forest to regenerate naturally is ensured.	X		X	X	X	
Representative areas, especially sites of ecological importance, are protected or appropriately managed.			X	X	X	
Corridors of unlogged forests are retained.			X	X		
No chemical contamination to food chains and ecosystem.	X		X	X		
Ecologically sensitive areas, especially buffer zones along watercourses, are protected.	X		X	X	X	
No inadvertent ponding or waterlogging as a result of forest management.	X			X		
Soil erosion is minimized.	X			X		
(Implied) Forest Management Maintains Fair Intergenerational Access to Resources and Economic Benefits						
Stakeholders'/forest actors' tenure and use rights are secure.		X	X	X	X	X
Tenure/use rights are well defined and upheld.	X	X	X	X	X	X
Forest-dependent people share in economic benefits of forest utilization.		X	X	X	X	
Opportunities exist for local people/forest-dependent people to get employment and training from forest company.		X	X	X		
(Implied) Stakeholders, Including Forest Actors, Have a Voice in Forest Management						
Stakeholders/local populations participate in forest management.		X	X	X	X	
Effective mechanisms exist for two-way communication related to forest management among stakeholders.		X	X	X		
Forest-dependent people and company officials understand each other's plans and interests.		X	X	X		
Forest-dependent people/stakeholders have the right to help monitor forest utilization.		X	X	X		
Conflicts are minimal or settled.		X	X	X	X	

ESSD FOREST POLICY IMPLEMENTATION REVIEW AND STRATEGY PROCESS: LIST OF ANALYTICAL STUDIES

Topic	Responsible party	Date
Challenges and recommendations: Contribution to the World Bank's Forest Policy Implementation Review and Strategy	IUCN/WWF	30 March 2000
Forest product market developments: the outlook for forest product markets to 2010 and the implications for improving management of the global forest estate	Adrian Whiteman, Christropher Brown, Gary Bull	July 1999
Beyond sustainable forest management: opportunities and challenges for improving forest management in the next millennium	C. Lennart, S. Ljungman, R. Michael Martin, Adrian Whiteman	December 1999
Towards sustainable forest management: an examination of the technical, economic and institutional feasibility of improving management of the global forest estate	Arnoldo Contreras-Hermosilla	July 1999
Forests and sustainable livelihoods: Current understandings, emerging issues and their implications for World Bank Forest Policy and funding priorities.	Gill Shepherd, Mike Arnold, and Steve Bass	September 1999
Forest carbon: A discussion brief on issues, project types, and implications for the World Bank's Forest Policy Strategy	DECRG/Kenneth Chomitz	May 1999
Corrupt and Illegal Activities in the Forestry Sector: Current understandings, and implications for World Bank Forest Policy	Debra J. Callister	May 1999
Indigenous peoples and forests: Main issues	Marcus Colchester	November 1999
Recent experience in collaborative forest management approaches: A review of key issues	Intercooperation/Jane Carter	May 1999
The World Bank and non-forest sector policies that affect forests	David Kaimowitz and Arild Angelsen	May 1999
Valuing forests: A review of methods and applications in developing countries	Joshua T. Bishop	July 1999
Plantations: potential and limitations	P. D. Hardcastle	October 1999
Notes on forestry legislation and enforcement	FAO	February 2000
Non-industrial private forest ownership/ privatization processes	Indufor Oy	November 1999
Institutional and legal framework for forest policies in ECA region and selected OECD countries—a comparative analysis	Birger Solberg and Kazimierz Rykowski	April 2000
Indigenous peoples, forestry management and biodiversity conservation	Jason W. Clay, Janis B. Alcorn, and John R. Butler	January 2000
Certification of forest management and labeling of forest products	Markku Simula and Indufor Oy	September 1999
The right conditions: The World Bank, structural adjustment, and forest policy reform.	Frances Seymour and Navroz Dubash	March 2000

ESSD REGIONAL CONSULTATIONS, FEBRUARY–MAY 2000

Region/country	Consultation dates	Location
Africa	3 to 5 May 2000	Johannesburg, South Africa
Brazil	15 to 16 March 2000	Brasilia
East Asia and Pacific	26 to 28 April 2000	Singapore
Europe and Central Asia	3 to 5 April 2000	Joensuu, Finland
Latin America and Caribbean	3 to 5 May 2000	Quito, Ecuador
North America	23 to 24 March 2000	Washington, D.C.
South Asia	17 to 19 April 2000	Rajendrapur, Bangladesh
Western Europe	10 to 11 April 2000	Zurich, Switzerland
Middle East and North Africa	23 to 25 February 2000	Tunis, Tunisia

ANNEX I: WORLD BANK OPERATIONAL POLICY 4.36, FORESTRY

September 1993
These policies were prepared for use by World Bank staff and are not necessarily a complete treatment of the subject.

Note: This document is based on *The Forest Sector: A World Bank Policy Paper*, 7/18/91, and also complements the following Bank guidelines: OD 4.01, Environmental Assessment; OD 4.20, Indigenous Peoples; OD 4.30, Involuntary Resettlement , and OMS 2.36, Environmental Aspects of Bank Work. Staff should also consult OD 14.70, Involving Nongovernmental Organizations in Bank-Supported Activities; OPN 11.02, Wildlands ; and OPN 11.03, Management of Cultural Property in Bank-Financed Projects. Questions may be addressed to the Director, Agriculture and Rural Development Department.

Bank ["Bank" includes IDA, and "loans" include credits] involvement in the forestry sector aims to reduce deforestation, enhance the environmental contribution of forested areas, promote afforestation, reduce poverty, and encourage economic development. In pursuit of these objectives, the Bank applies the following policies:

(a) The Bank does not finance commercial logging operations or the purchase of logging equipment for use in primary tropical moist forest. In borrowing countries where logging is being done in such forests, the Bank seeks the government's commitment to move toward sustainable management of those forests, as described in para. 1(d) below, and to retain as much effective forest cover as possible. Where the government has made this commitment, the Bank may finance improvements in the planning, monitoring, and field control of forestry operations to maximize the capability of responsible agencies to carry out the sustainable management of the resource.

(b) The Bank uses a sectorwide approach to forestry and conservation work in order to address policy and institutional issues and to integrate forestry and forest conservation projects with initiatives in other sectors and with macroeconomic objectives.

(c) The Bank involves the private sector and local people in forestry and conservation management or in alternative income-generating activities. The Bank requires borrowers to identify and consult the interest groups involved in a particular forest area.

(d) The Bank's lending operations in the forest sector are conditional on government commitment to undertake sustainable management and conservation-oriented forestry. Such a commitment (which may be reflected in specific conditionalities; see Good Practices 4.36 for examples) requires a client country to:

(i) adopt policies and a legal and institutional framework to (a) ensure conservation and sustainable management of existing forests, and (b) promote active participation of local people and the private sector in the long-term sustainable management of natural forests (see paras. 19–20 of OD 4.01, Environmental Assessment);
(ii) adopt a comprehensive and environmentally sound forestry conservation and development plan that clearly defines the roles and rights of the government, the private sector, and local people (including forest dwellers) (see OD 4.20, Indigenous Peoples);
(iii) undertake social, economic, and environmental assessments of forests being considered for commercial use;
(iv) set aside adequate compensatory preservation forests to protect and conserve biological diversity and environmental services and to safeguard the interests of forest dwellers, specifically their rights of access to and use of designated forest areas; and
(v) establish institutional capacity to implement and enforce these commitments.

(e) The Bank distinguishes investment projects that are exclusively environmentally protective (e.g., management of protected areas or reforestation of degraded watersheds) or supportive of small farmers (e.g., farm and community forestry) from all other forestry operations. Projects in this limited group may be appraised on the basis of their own social, economic, and environmental merits. However, they may be pursued only where broad sectoral reforms are in hand, or where remaining forest cover in the client country is so limited that preserving it in its entirety is the agreed course of action.

(f) In forest areas of high ecological value, the Bank finances only preservation and light, nonextractive use of forest resources. In areas where retaining the natural forest cover and the associated soil, water, biological diversity, and carbon sequestration values is the object, the Bank may finance controlled sustained-yield forest management. The Bank finances plantations only on nonforested areas (including previously planted areas) or on heavily degraded forestland.

The Bank does not finance projects that contravene applicable international environmental agreements.

Annex A to Operational Policy 4.36, Forestry:
The following definitions apply in this statement:

(a) *Primary forest* is defined as relatively intact forest that has been essentially unmodified by human activity for the previous 60 to 80 years.
(b) *Tropical moist forest* is generally defined as forest in areas that receive not less than 100 mm of rain in any month for two out of three years and have an annual mean temperature of 24°C or higher. Also included in this category, however, are some forests (especially in Africa) where dry periods are longer but high cloud cover causes reduced evapotranspiration.
(c) *Carbon sequestration* refers to the process whereby forested areas retain a revolving but stable store of organic carbon in their biomass. Clearing, burning, or otherwise substantially altering the forest increases the net release into the atmosphere of carbon-based gases that contribute to the greenhouse effect.
(d) The term *local people* describes the broad group of people living in or near a forest, with some significant level of dependence upon it. The term includes forest dwellers, indigenous forest-adjacent populations, and recent immigrants.
(e) *Sustainable management* of natural forests means controlled utilization of the resource to produce wood and nonwood benefits into perpetuity, with the basic objectives of long-term maintenance of forest cover and appropriate reservation of areas for biodiversity protection and other ecological purposes.
(f) A *natural forest* is an area in which the cover has evolved naturally so as to provide significant economic and/or ecological benefits, or one that is sufficiently advanced in regeneration and recovery from disturbance as to be judged in near-natural condition.

ANNEX J: WORLD BANK GOOD PRACTICES 4.36, FORESTRY

May 1993

Introduction

This statement gives guidelines[1] for implementing the Bank's operational policy on the forest sector.

Bank[2] lending in the forest sector emphasizes the development of forest resources to provide for a sustainable stream of direct or indirect benefits to alleviate poverty, advance the status of women, and enhance community income and environmental protection. Bank forest sector policy now recognizes that, in most cases, progress across this range of objectives requires a program approach to the sector, rather than a discrete project approach. Ongoing investments and existing forest policies, often developed piecemeal, tend to involve excessive public intervention in the sector. A program approach requires a consistent sector policy that builds on favorable macroeconomic policies, promotes links between the forest sector program and project design, and recognizes the relationships among forest, people, and culture. The program approach therefore requires (a) sufficient Bank involvement in the forest sector to allow a dialogue on sector strategy and policy reform; (b) strong linkages between the forest sector program and project design and the broader macroeconomic dialogue; and (c) involvement of a wide range of government agencies and interested parties outside of government to ensure broad consensus on reform priorities.

Country Economic and Sector Work

It is essential to the sectorwide strategy for forestry development that country economic and sector work (CESW) be as highly developed as possible prior to the start of major lending in the sector. Alternatively, when resource constraints are severe, CESW may be carried out during the early stages of project identification and preparation. In such cases, special efforts are required to ensure that analysis of sector issues is not lost among project design concerns. Bank staff should obtain explicit recognition of the joint policy reform and investment preparation effort from government, and especially from their counterparts in coordinating agencies outside the forest sector (e.g., ministry of finance, planning commission).

Economic work

The Bank's country economic work (especially the country economic memorandum) assesses the macroeconomic policies likely to affect the forest sector. The forest sector policies advocated during project preparation will grow out of those put forward in the country economic memorandum, the national Environmental Action Plan, and the Consultative Group process[3] (where Consultative Groups are an appropriate mechanism), and as structural or sector adjustment inputs.[4]

CESW gives special attention to the macroeconomic policies that are likely to affect resource use and the environment. In this category are policies governing:

(a) administered resource pricing, in which the government influences the flow of resources by using nonmarket interventions to promote domestic processing or similar objectives;
(b) revenue sharing among different levels of government and among public and private sectors and local communities;
(c) criteria used to identify and appraise public investments in agriculture and infrastructure that affect the management of forestland;
(d) institutional budgets and funding procedures, and the general incentives for rent seeking in the economy;
(e) general trade and industry, especially policies that determine which exportable natural resources are to be processed domestically;
(f) energy, especially the potential for fuelwood substitution and for improving efficiency of energy use where nonsustainable fuelwood gathering is a serious problem; and
(g) the incentive framework for channeling investment resources to the most efficient users of funds, especially the private sector, including small farms, communities, tribes, and women's groups.

Sector work

Forest sector lending requires a strong sector work basis. Sector work in advance of detailed project design helps integrate forest sector aims and objectives with wider economic and environmental concerns. The sectoral topics covered vary from one country to another. The policy areas that may need to be investigated include the following:

(a) *Resource pricing and methods of sale.* Measures related to the recovery of economic rent and to revenue collection mechanisms can be incentives to observe environmental and management prescriptions—for example, (i) using performance bonds rather than difficult-to-enforce penalty clauses; (ii) using resource assessment rather than extracted volume measurement to encourage efficiency in operations; and (iii) incorporating in the concession agreement requirements to finance specified nature conservation and community involvement activities. Sector work should aim to quantify the effect of prices on rates and on levels of forest use, the distributional impacts of price changes, and the effect of pricing mechanisms on public and private sector risk bearing.

(b) *Forest industry policies.* Policies of state intervention to stimulate the development of industry based on forest resources or to increase value added in the forest sector should be assessed. Quotas, tariffs, and overvalued foreign exchange rates on forest product imports can artificially increase private financial returns on forest exploitation, establishing incentives to exploit forest resources more than is economically efficient or physically sustainable. Conversely, log export bans and taxes can reduce the private financial value of forestry activities below their economic value, leading to inefficiencies in log use and reduced investment in forestry. Where appropriate, alternative policies to encourage the growth of efficient and competitive value-added operations and to achieve equity objectives should be presented.

(c) *Forest resource information.* Sector work should assess the government's database, procedures for managing and analyzing data, inventory programs, and the adequacy of public access to information, including data on forest product prices. Recent studies emphasize the economic importance, particularly to the poor, of nontimber forest products (e.g., nuts, fruits, and medicinal plants), but information on these products is rarely available or analyzed. The Bank encourages governments to develop resource monitoring systems (including natural resource accounting).

(d) *Human resource development.* Sector work should assess (i) the availability of skills in both government and the private sector, and (ii) the adequacy of personnel policies (including staff rotation systems and opportunities for specialization), compensation and field allowances, and the systems for delivering training to farmers and forest workers.

(e) *Research.* Sector work should assess the forestry research program, focusing especially on the balance of research between and among various areas—natural and man-made forests, forest management and forest products (including nontraditional forest products), social and economic aspects, and ecological and technical areas. The Bank encourages the development of international research links with member institutions of the Consultative Group on International Agricultural Research, the International Union of Forestry Research Organizations, and national forestry research systems.

(f) *Resource mobilization.* Sector work should assess the adequacy and efficiency of funding mechanisms for public and private forestry operations, including credit arrangements and earmarking of forestry revenues. Such mechanisms should ensure the availability of funds for regeneration, forest protection, and investment in other sectors of the economy.

(g) *Participation and role of private sector.* Sector work should examine the division of responsibility among the different levels of government, parastatals, local communities, and the private sector. It should consider the scope for devolving responsibility for land management and directly productive investment to local communities and private firms while focusing government efforts on regulations and technical assistance.

(h) *Environmental framework.* Sector work should examine the legal and institutional basis for ensuring environmentally sustainable development of the forest sector. The government should have in place adequate provisions for conserving protected areas and critical watersheds and for establishing environmental guidelines and monitoring procedures.

Project Processing

No special procedures are required for the processing of forestry projects. However, forestry projects, especially those pursuing broad program objectives, present special analytic issues that need to be taken into consideration throughout the project cycle. Normally, although the borrower "owns" the reform program and is responsible for preparation and implementation, the Bank needs to supply guidance and support for communication among government agencies and nongovernmental organizations (NGOs).

Identification and preparation
As has been indicated, Bank projects normally evolve from a program of CESW and an agreed reform agenda. In the absence of completed CESW, the identification process should provide for assessment of the fundamental policy issues facing the sector. Projects should aim to enhance environmentally sustainable development while taking into account limits on absorptive capacity and management control.

The Bank may consider financing a broad range of investment activities in or related to the forest sector. Several possible areas for investment are described below.

Preservation and management of intact forest areas
The Bank may support initiatives to expand forest areas allocated as parks and reserves and to institute effective management and enforcement in new and existing areas. The Bank stresses approaches to management of protected areas that consider the welfare of forest-dwelling people and incorporate local people into protection, benefit sharing, and planning.

In tropical moist forests, the Bank adopts, and encourages governments to adopt, a precautionary policy toward use. The Bank emphasizes support to programs that involve institutional development, forest protection measures, and nonforest income-generating projects that aim primarily to preserve tropical moist forests. The Bank makes special efforts to support economic development in poor, densely populated areas around such forests or in forest encroachers' areas of origin. The Bank also supports ameliorating environmental damage in temperate and boreal forests, directing its investments toward programs to rehabilitate and reforest degraded forestland and to reduce industrial pollution and conserve energy.

Resource expansion and management intensification
The Bank may finance projects to create additional forest resources or to expand and intensify the management of areas suitable for sustainable production of forest products. The Bank promotes rural people's participation in tree planting and conservation of indigenous woodlands.

The key to increasing forest investment is a balance among economic incentives, security of tenure, motivation, and technical assistance. The Bank directs special efforts toward agroforestry technologies that can improve soil fertility, conserve soil moisture, and increase crop and livestock yields. It may encourage cash crop tree farming in rural areas where such farming does not impair local people's access to essential fuelwood and fodder supplies. The Bank emphasizes developing market intelligence and marketing systems for cash crop tree farming and for assisting small-scale wood-using enterprises in rural areas. To reduce pressure on the existing forest resource base, the Bank may support the establishment of plantations outside areas of intact forest, where such activity is socially, environmentally, and economically acceptable. The primary target areas for new planting are potentially productive degraded forests, wastelands, forest fallows, shrublands, and abandoned farmlands. The interests of communities that depend on such areas must be considered in setting target areas.

Institutional reform and strengthening
The Bank recognizes the critical need to restructure forestry institutions, improve training and equipment, and introduce greater accountability and higher performance standards into the public sector. It may support the use of private sector contractors and consultants as auditors and monitors, and more rigorous intersectoral oversight by ministries of agriculture, environment, planning, finance, and so forth.

Forestry research and development
The Bank may support forestry research and technology development, both as project components and as free-standing projects. Provision for dissemination of results and for technology transfer should be made during preparation.

Improved processing and demand reduction
The Bank may support direct interventions to encourage conservation and the use of more efficient technologies, including research and training to improve the fuel efficiency of household stoves. In addition, the Bank may support research and investment in new technologies to reduce industrial demand for wood and to better match industrial demand with available resources—for example, retooling and upgrading to permit processing plants to use small logs and wood residues.

Resource assessment
The Bank may support forest resource assessments through surveys, inventories, and mapping. Emphasis should be given to the application of Geographic

Information Systems and other recent advances in information technology for project planning, monitoring, and evaluation.

Environmental assessment
Bank staff ensure that the borrower conducts an environmental assessment (see OD 4.01, *Environmental Assessment*).

Appraisal
During appraisal the Bank ensures the overall viability of the proposed project. In particular, appraisal of forestry projects needs to focus on five areas: economic and policy framework, institutional arrangements, technology, environmental protection, and local participation.

Economic and policy framework
Appraisal should ensure that the project is consistent with the policy reform priorities identified through CESW and that adequate provisions are made for marketing (particularly by the rural poor) and procurement. In particular, the Bank must ensure that procedures for land acquisition and forestland designation reduce inappropriate clearing for agriculture and that pressures from grazing and other uses are taken into account. The Bank should ensure that the borrower is committed to a policy of sustained resource management through the application of scientific forest management systems.

Institutional arrangements
Appraisal establishes that all policy initiatives have an identified legal and regulatory basis; that agencies have the capability to fund, review, and monitor—and enforce compliance with—the planned initiatives; and that interagency responsibilities and coordinating mechanisms are established. The Bank ensures that the roles of the public and private sectors in project implementation are clearly defined.

Technology to increase productivity
Projects should employ high-quality technology appropriate to local conditions and comparable to best practice in other agricultural subsectors such as tree crops. The Bank seeks to ensure that technological improvements introduced by projects are disseminated throughout the sector.

Environmental protection
Environmental guidelines for forest management should provide for adequate attention to soil and moisture conservation; protection of watersheds, streams, and waterways; maintenance of adequate areas to conserve biodiversity; species mix in plantations; and pest and fire control.

Local participation
Local communities and NGOs should participate in project design, and the project's legal and financial mechanisms should be adequate to secure their participation in the project. The Bank considers proposed staffing patterns to ensure that field staff have appropriate skills and training to work effectively with local people. For example, since women are usually the main collectors of fuelwood and forest products, especially for domestic use, women may be especially effective as extension workers and as workers in such activities as nursery management and seedling distribution.

ANNEX K: IMPLEMENTATION OF THE 1991 FOREST STRATEGY IN THE IFC'S PROJECTS

The following document, published here with permission, is the Executive Summary of the report prepared by the Operations Evaluation Group of the IFC for this review.

Background

In 1991, the Bank issued a forest paper[1] (1991 forest paper) that set out its strategy for intervention in the forest sector and other sectors affecting forests. The forest paper focused special attention on the conservation of primary tropical moist forests (TMFs), and its objectives were to reduce deforestation and increase planting of trees in order to expand forest cover. In 1993, the Bank issued Operational Policy 4.36[2] (Bank's OP 4.36) that reflected the policy content of the strategy and introduced Good Practices 4.36[3] (GP 4.36) to provide staff with direction on implementing the strategy.

Since the 1991 forest paper was written primarily for Bank and IDA operations, it provided limited and unclear guidance for IFC operations. In recognition of this fact, the IFC issued a memorandum to its Board[4] (1991 forest memorandum) to clarify how the 1991 forest paper would be applied to its operations, including a commitment that all the IFC's projects would conform with the "spirit and intent" of the 1991 forest paper. When the Bank's OP 4.36 was introduced in 1993, the IFC adopted it automatically for its forest operations. The 1991 forest paper, the 1991 forest memorandum, and the Bank's OP 4.36 are collectively referenced as the "IFC forest strategy" in this paper, and implementation of the strategy in the IFC's operations is the subject-matter of this evaluation.

This study is based on (i) a review of all the forest-based investments approved by the IFC during FY85–98 to identify the changes induced in IFC operations on a "before/after strategy" basis; (ii) a review of selected non-forest infrastructure projects supported by the IFC in FY92–98 which potentially have impacts on forests; (iii) a review of selected financial intermediary investments approved in FY92–98 to assess evidence of adherence by the intermediaries and their IFC-funded sub-projects to the requirements of the forest strategy; and (iv) case studies on a sample of 14 forest-based companies visited to assess their on-the-ground results from the perspective of the 1991 Forest Strategy's objectives. The sampled projects in the case studies were broadly representative of the regional distribution of the IFC's portfolio at the end of FY98 of forest-based projects that relied on local resources.

In the context of the WBG's review of its forest strategy and policy, the purpose of this study is to answer the following questions: (a) how effective was the 1991 Forest Strategy? (b) what were its impacts on the IFC's projects? (b) how adequate was the IFC's implementation of the strategy in its operations? and (c) what are the lessons from the IFC's experience to guide strategy revision and future operations? In addition, the study reviews the IFC's OP 4.36, approved in 1998, to establish its coherence with the forest strategy.

This study complements the Operations Evaluation Department's (OED's) evaluation of the Bank's experience under its forest strategy.

Main Findings

Effectiveness of strategy

Based on the findings of the portfolio review and case studies, the intention of the forest strategy to engage the private sector in sustainable forest management is not attuned to realities on the ground. Part of the reason has to do with economics: over half of IFC projects approved since 1991 use government-owned forests, and the stumpage paid in some of the cases studied is lower than the real cost of managing the forests sustainably. Hence the private concessionaires have a financial disincentive to encourage the more costly option of sustainable forest management. Part of the reason has to do with the manner in which forest concessions are operated. A concession that is not fixed to a specified area in the forest does not put any part of the forest under the control of individual private operators. In the absence of such delimited areas of control, private operators will generally not accept contractual responsibility for sustainable management of the forest. Part of the reason has to do with ownership of the forest: in general, private operators are not given the contractual right to manage sustainably forests that belong to government. The case studies found that governments want to retain ownership and management control of the forestlands for varying reasons. In Sub-Saharan Africa, land tenure arrangements discourage or prohibit ownership by foreigners, whereas many of the forest-based companies are controlled by foreign investors. In one Asian country, the forests belong to farm collectives organized under the country's political structure and are not available for foreign ownership. In Eastern Europe, forests are considered to be part of national endowments that should be preserved in the hands of the state. Reinforcing the status quo within these cultural and legal frameworks are powerful vested interests, who in many cases are enabled by the lack of transparency and accountability to extract private gain from the state's control of the forest resources.

The forest strategy implied a commitment by the IFC project companies to work with the local people, interest groups and forest dwellers in forest areas, and that has proven to be unrealistic. Lacking ownership or contractual control over the forests, private operators do not welcome— and have no incentive to accept— this responsibility and the strings attached in the form of entanglement in local law enforcement and politics.

Impact of the strategy

The 1991 Forest Strategy did not induce reforestation and increased tree planting in IFC projects. The projects that engage in tree planting need to do so for commercial reasons, to create a sustainable and economically optimum raw material resource for their operations. Similarly, IFC projects that reforest harvested areas do so to ensure the long-term sustainability of their capital-intensive manufacturing operations.

However, the 1991 forest paper created a heightened awareness of the environmental sensitivity and value of natural forests as ecosystems, particularly TMFs, and has led to fundamental changes in project selection, processing, and monitoring by IFC. The forest sector also featured importantly as a driver of the steps taken by the IFC during the past three years to strengthen its policies, procedures and staff resources to meet environmental requirements.

The focus of the forest strategy on TMFs neglected other forest types with differing biodiversity situations and issues. In particular, the strategy did not provide guidance specifically relevant to projects utilizing temperate and boreal forests, which comprised all of the IFC's forest-based operations that were approved in the post-strategy period.

The strategy's strong emphasis on the ban on commercial logging in the TMFs had a chilling effect on the IFC. In order to avoid any association with deforestation, the IFC made a conscious decision to screen out this type of operation completely, and subsequently turned down several proposals submitted to it. Thus, those proposals were denied the valuable contribution the IFC could have made through its at-entry structuring role, including measures to ensure sustainable resource and environmental management. The study could not ascertain whether any of these projects were implemented without IFC financing.

Strategy implementation

Overall, the IFC's operations approved since FY92 have been consistent with the twin objectives of the forest strategy of abstaining from activities which add to the loss of TMFs and encouraging reforestation. The IFC has diligently upheld the prohibition against financing commercial logging in TMFs, and in its mainstream operations it has not approved a single forest-based investment in this category since the strategy became operational. However, two small investments—approved under streamlined review procedures through the IFC's small- and medium-size enterprise (SME) facilities—financed a company engaged in transporting logs harvested under concessions in TMFs. While these investments were consistent with the letter of the forest strategy, they digressed from its "spirit and intent," and the environmental review procedures in place at the time failed to detect the inconsistency. In addition, two other small investments also approved under the facilities were in companies whose non-project operations, or those of their sponsors, involved logging in TMFs.

The IFC has contributed to the establishment of project-owned large plantations, particularly in Latin America and Asia, which comprise about 40 percent of post-strategy approvals and are managed sustainably. Even in countries where the forests are owned by governments, IFC staff ensured that forest operations that supplied IFC projects were carried out in a sustainable manner and in accordance with good industry practice. The IFC has also supported several projects that rely on wood wastes, wastepaper and other recycled materials, thus contributing to conservation of forest resources.

In addition, the IFC has undertaken special initiatives that supported forest conservation and sustainable management. In 1991, the IFC sold its 58,000 hectares parcel of land in Paraguay (acquired in a foreclosure) for the creation of a nature reserve in perpetuity. Since 1995, the IFC has been administering a $16 million grant from the Global Environmental Facility (GEF) to support SMEs promoting climate change initiatives and sustainable use/conservation of biodiversity.

In its financial intermediary (FI) investments, the IFC has not generally required its borrowers to report on the sub-projects they assisted and, as a result, IFC does not know about the possible involvement of these sub-projects in the forest sector. On the basis of a file review, the study could not ascertain the extent to which IFC-supported FIs and their sub-projects in countries with threatened TMF are conducting their operations in a manner consistent with the forest strategy. Since 1994 IFC has been training the FIs on environmental risk management, including sessions on the IFC's forest strategy and its prohibition against financing commercial logging in the TMFs. In the IFC-specific environmental and social safeguard policies that were approved by the Board in July 1998 and which define the current framework, special requirements were developed for FI projects, and environmental requirements and review procedures were specified for sub-projects according to a tier classification.

However, there have been process weaknesses in the implementation of the 1991 Forest Strategy by IFC. Specifically, the IFC did not (a) disseminate the forest strategy to staff to ensure their familiarity with the objectives and relevant requirements; (b) reflect the strategy in its operations manual or provide staff with a manual similar to the Bank's GP 4.36; (c) put clear screening/appraisal guidelines in place for processing forest-impacting projects; nor (d) for relevant projects, include explicit covenants in the investment agreements to commit its project companies to their expected responsibility under the IFC forest strategy.

Forest certification

Forest certification, which operates as an association of forest areas certified to be sustainably managed, can promote the aims of the forest strategy because it enables consumers to recognize products from such forests. However, certification has tended to serve as a marketing tool for exports and is, thus, not relevant for many forest-based companies whose products sell in the domestic markets or trade internationally as commodities. In fact, certification is unpopular among forest-based companies that do not own their forest resources and are not responsible for management of the forests. The recent proliferation of certification schemes is creating an unhealthy rivalry for "membership recruitment." IFC does not insist on forest certification but encourages its projects to obtain it.

IFC's new forest paper

IFC's OP 4.36 was introduced in July 1998 and, by design, was "harmonized" with the Bank's version. Parts of the document are vague and can be misinterpreted unless the reader is aware of and acquires a good understanding of the 1991 forest paper. But a major shortcoming of the IFC's OP 4.36 is that it does not contain any reference to the forest paper, unlike the Bank's 1993 version from which it was derived, and the source and objectives behind its policy content are therefore missing.

ANNEX L: RECENT WORLD BANK REVIEWS OF FOREST POLICY IMPLEMENTATION

The 1994 Agriculture Department Review

This review acknowledged considerable progress but concluded that deforestation continues apace in developing countries, forests remain seriously undervalued, and interventions in the sector remain controversial.

It recommended four priorities:

- *Target poverty reduction.* Help countries use forest resources to reduce rural poverty based on a participatory approach.
- *Reconcile conservation and forest use.* Stimulate government commitment to sustainable and conservation-oriented forest management while avoiding financing commercial logging in primary moist tropical forests.
- *Support policy and institutional reforms to promote private investment.* Conduct those reforms in a manner consistent with the objectives of poverty reduction and environmental protection.
- *Pay more attention to the impact of nonforest activities on forests.* In particular, incorporate intersectoral linkages in country assistance strategies and country economic and sector work.

The 1996 Agriculture Department/OAG Review

This review identified generic implementation issues in 16 projects and developed strategies and approaches for dealing with them. It identified two fundamental causes of poor performance in the forest sector: inadequate borrower commitment to the necessary policy and institutional reforms and inadequate project design.

The review recommended that the Bank:

- Review policies and procedures to ensure that priorities are set in accordance with the importance of the resources involved and the tractability of the problem.
- Establish a clear link between forest issues and country assistance strategy and structural adjustment loan processes and broaden the dialogue to include civil society.
- Develop a strong knowledge base to underpin this strategy.
- Redeploy scarce technical and operational staff familiar with forest issues flexibly and judiciously.
- Develop new action-oriented partnerships with the private sector and donor and NGO communities.

ANNEX M: REPORT FROM COMMITTEE ON DEVELOPMENT EFFECTIVENESS (CODE)

OED: A Review of the World Bank's 1991 Forest Strategy and Its Implementation, and OEG Review—Implementation of the 1991 Forest Strategy and IFC's Projects

The Committee met on December 23, 1999 to discuss the OED preliminary report *A Review of the World Bank's 1991 Forest Policy and Its Implementation* (CODE99 96), prior to the Report's consultations with external stakeholders. A Green Sheet was issued (CODE99-101). The Committee met again on July 10, 2000 to consider the final report (CODE2000-70) which incorporate the results of the GEF and IFC reviews and of external consultations, and clarifies all statements regarding policy. The Committee also discussed the *OEG Review: Implementation of the World Bank's 1991 Forest Policy in IFC's Projects* (CODE2000-67); the *IFC Management Response* (CODE2000-73); and ESSD's "Update on Preparation of the Forest Strategy," which provides an update on the development of Management's forest strategy and outlines some of the issues and directions that are emerging from the process so far.

Following the CODE Report on both meetings.

The Committee welcomed the OED report and appreciated the highly participatory and comprehensive nature of the report and the widespread external consultations that had taken place. The Committee agreed that although the Bank broke new ground in 1991 by emphasizing conservation—a laudable goal—Bank involvement in the sector since then has been conditioned mainly by "safeguard" issues, i.e. the ban on Bank financing of logging, protection of the environment associated with infrastructure projects and of indigenous people. This produced a chilling effect and led to less proactive Bank involvement than would have been desirable. As a result, the Bank had had little effect in terms of reducing the rate of deforestation, significantly improving forest cover or protecting biodiversity. It welcomed the multisectoral and multistakeholder approach proposed by OED.

The Committee focused on the following major issues:

Bank's Role: Many directors supported the Bank playing a more active, synergetic role in the forest sector at both the country and global levels in view of the importance of the sector for poverty reduction, the global public goods dimensions, and the significant value added that the Bank has provided in cases where it has a substantial involvement in the forest sector. However, others asked whether the Bank had a comparative advantage in this sector, and commented that given its financial and institutional limitations, and reduced internal capacity, the Bank Group should be selective in its objectives and its role. The Committee agreed that it would be important to continue this discussion in the context of reviewing the forthcoming forest sector strategy.

Synergy between Development and Conservation: The Committee recognized the critical relationship between development and conservation. It therefore supported the need to broaden the strategy to include country concerns such as poverty reduction, energy use and substitution and incentives for sustained forest-management at the national and individual level. In this context, the Committee agreed that conservation should be broadened to encompass all natural forest types and that both indigenous and other local people, especially the poor, should be the focus of Bank interventions, and congratulated OED for the treatment of gender issues.

Governance and Capacity Building: Committee members supported the focus on more effective country enforcement of logging restrictions and improved governance, and agreed that the Bank should support these activities within an overall framework that includes sound forest management and poverty reduction. Some members also endorsed the point made in the Report that, to be effective, forest strategies must fit the specific geographical, biophysical, demographic, social, cultural and economic circumstances for which forest interventions are designed and called attention to the implications of this point to the design of a new Bank policy in this area.

Financing Mechanisms: The Committee recognized the importance of concessional and grant financing for the forest sector in view of the long-term and risky nature of many forest sector investments. While some members supported the idea, others expressed reservations about the proposal for a global compensation mechanism to pay countries simply for positive international externalities of forest conservation, funded with consessional resources. They were also concerned that such a mechanism would be difficult to implement, that compensating countries or people just for keeping forests could create an incentive for rent seeking, that it would be ineffective in the absence of local capacity to enforce compliance with logging restrictions and redundant where such capacity exists.

Bank's Organization and Instruments: Many members agreed that if the Bank is to play a larger role, it should build up a critical mass of skills and resources, improve internal coordination, develop a better monitoring system on compliance with safeguards, and promote a holistic approach through increased treatment of forest issues in CASs, CDFs, and ESW. The Committee agreed that conditionalities in adjustment loans were not always sufficient in promoting effective forest sector reforms and could set back national ownership of reforms. There was a need to explore further the balance of lending to forest sector in comparison to lending for forest policy in sector and adjustment loans, and lending for forest as a component of agricultural or other types of investments including APLs. Some members noted that IFC's work depended on economic and financial incentives for the private sector, and, that while sharing the objectives for sustainable forest development, different approaches may be needed.

Compliance with Global Agreements and International Processes: The Committee stressed the need for effective coordination with other global partners, such as the Bio-diversity Convention, Food and Agricultural Organization, the U.N. International Forum on Forests, World Wide Fund for Nature, Interagency Task Force on Forests, as well as with the emerging framework of the Kyoto Protocol which does not allow for carbon sequestration activities in developing countries.

IFC and Forestry: The Committee welcomed the OEG review and appreciated recent efforts to increase awareness of IFC staff. An important finding of the review was that IFC in its mainstream operations had upheld the forest strategy's prohibition against financing commercial logging in tropical moist forests. In addition, within IFC environmental awareness had been heightened and project environmental review procedures and policies had been strengthened. Other findings included the lack of consensus within IFC regarding the applicability of the 1991 forest paper to its operations; the lack of definition of key terms in the 1991 forest memorandum; and the difficulty in realizing the goals of the 1991 forest paper on the ground. As a result the impact of the forest strategy had been limited and several shortcomings would need to be addressed in the revised strategy, including more clarification on financial intermediaries; increasing awareness about the strategy among IFC staff and client companies; addressing the needs of SMEs; and improving tracking compliance in projects of financial intermediaries. Management looked forward to a revised strategy and noted that OEG's findings and recommendations will have major implications for its work.

Development of New Bank Group Strategy: Members agreed with OED's recommendations for the design of a new Bank Group forest strategy, and welcomed the broadening of the discussion to include the IFC, MIGA and GEF. In this context, speakers noted that the Bank Group's role in forestry was, to a large extent, unrealized, and looked forward to the development of a new strategy. Speakers generally agreed that the Bank Group should rely on effective partnerships and its convening power, and, at the same time, develop clear proposals for moving forward in certain areas such as illegal logging, forestry management and safeguards. Speakers proposed that the new forest strategy examine how forest issues could be mainstreamed into Bank work such as CASs, CDF and APLs and other lending instruments.

Next Steps: OED and OEG will proceed with publication of the final reports, taking into account the comments made at both the meetings. Management would continue to brief CODE on concrete issues relating to the development of the Bank Group strategy, with a target completion date of December 2000.

Jan Piercy
Chairperson

ENDNOTES

Chapter 1

1. In this paper "1991 Forest Strategy" refers to the World Bank paper on forests (World Bank 1991a), Operational Policy (OP) 4.36, and Good Practices (GP) 4.36 (see box 1.1).

2. An OED evaluation of approaches to the environment in Brazil (World Bank 1992b) and a 1989 audit of a forestry project in Côte d'Ivoire (Loan 1735) raised several environmental concerns.

3. The 20 countries are Bolivia, Brazil, Cameroon, Central African Republic, Colombia, Congo Democratic Republic, Congo Republic, Côte d'Ivoire, Ecuador, Gabon, India, Indonesia, Madagascar, Malaysia, Mexico, Myanmar, Papua New Guinea, Peru, Philippines, and Venezuela.

4. Safeguard policies are directives to ensure that Bank lending operations do not harm people or the environment. In addition to Forestry (OP 4.36), these policies are Environmental Assessment (OP 4.01), Natural Habitats (OP 4.04), Pest Management (OP 4.09), Involuntary Resettlement (OD 4.30), Indigenous Peoples (OD 4.20), Cultural Property (OPN 11.03), Safety of Dams (OP 4.37), Projects in International Waterways (OP 7.50), and Projects in Disputed Areas (OP 7.60).

5. Some of the external benefits from forests—for example, reducing soil erosion, desertification, or degradation of watersheds—accrue to the countries in which the forests are located. Other benefits spill over national boundaries and affect the international community by reducing the rate of loss of biological diversity and the rate of change in the global climate, for example.

6. For the Brazilian Amazon, the Bank's 1991 forest sector paper did acknowledge an estimated value of carbon in undisturbed forests of $375 to $1,625 per hectare, while current land prices ranged between $20 and $300 per hectare. The benefit of saving 21 million hectares of forest ranged from $750 million to $3.2 billion, whereas the cost of acquiring the forest ranged from $420 million to $600 million (Schneider 1992). This divergence in costs and benefits at different levels is borne out in subsequent economic and sector work in Brazil.

7. World Bank workshop on sustaining tropical forests, sponsored by the Environment and Rural Development departments, held at Graves Mountain Lodge, Syria, Virginia, 2–7 October 1998.

8. Using the pattern of forest development first articulated by von Thunen, Hyde argues that the relative levels of forest value, agricultural land value, and protection costs will determine levels of secure agriculture, managed forestry, open-access agriculture, degradation, and deforestation. Consequently, countries would go through several stages of forest exploitation, including open-access agriculture, permanent agriculture, managed forestry, and enforceable conservation.

9. A recent national opinion poll on proposed changes to the Forest Code, released by the Brazilian environmental organization Instituto Socioambiental (Socioenvironmental Institute) and partner groups, found that, of the 503 people interviewed, 88 percent believe that the protection of Brazil's forests should increase, not decrease, and 90 percent believe that increasing deforestation in the Amazon to establish agricultural lands will probably not reduce hunger. When questioned about the restoration of fragile areas to prevent siltation, landslides, or floods, 87 percent stated they believe that property owners who deforest should be fined and forced to restore the vegetation.

Chapter 2

1. The OED Brazil forest sector study found the best treatment of intersectoral issues, but much of that work was carried out in the early 1990s. A recent piece of ESW in Brazil is also of high quality. The India country study noted that, despite two decades of lending, forest sector work has not addressed key challenges in the forest sector such as resource mobilization and reform of the forest departments. In China, where the Bank has had the largest lending program in forestry, there has been no sector work.

2. See Annex tables C.6 and C.7 for a complete breakdown of regional forest commitments by forest and forest-component projects.

3. As this report was being completed, the form for Project Status Reports was revised to include a section on "Compliance with Safeguard Policies," a welcome development in line with OED's recommendation.

4. Developments in Cameroon since preparation of the preliminary OED report are said to have grown more promising.

5. Adjustment lending accounts for 37 percent of total lending in the Africa Region (see Africa portfolio review, Kumar and others 2000c, and Cameroon country study, Nssah and Gockowski 1999, for details).

6. In an earlier review of Structural and Sector Adjustment for 1980–92, OED observed that adjustment measures, which are designed to have economic effects, may also affect the environment through the reallocation of the resources they engender. This effect may be positive or negative.

7. This section is based on analysis in Lele and others 2000d. See also Seymour and Dubash 2000.

8. FAO data show that of the estimated 1.3 billion people living in poverty, more than 70 percent are women. The number of rural women living in absolute poverty has risen almost 50 percent over the past two decades.

9. The review examined compliance with four safeguards for the forest and forest-component projects: OP 4.36, Forestry; OP 4.04, Natural Habitats; OP 4.01, Environmental Assessment; and OD 4.20, Indigenous Peoples. Projects with a potential impact on forests in the transportation, agriculture, electric power and energy, and mining sectors were examined for compliance with three safeguards: OP 4.36, Forestry; OP 4.04, Natural Habitats; and OP 4.01, Environmental Assessment.

10. The Bank has begun to use external stakeholders to monitor the implementation of safeguards. The Chad-Cameroon Pipeline, recently approved for Bank financing, has established an advisory group of independent international experts to monitor the project's social and environmental safeguards. The advisory group findings will be discussed by the senior management and Board of the Bank Group and then made public.

11. Results of workshop discussions in Brazil and China.

12. In several countries, small enterprises in the informal sector process large quantities of forest products, including timber. Improving the investment climate for the private sector as a whole will require due consideration of the needs of these small enterprises.

13. Biodiversity loss is a major issue in several countries, but it is not clear whether it is high on the agenda of governments that face many other developmental challenges (see Africa portfolio review, Kumar and others 2000c).

Chapter 3

1. This section is based on the six country studies carried out by OED and the outcomes of discussions involving a range of stakeholders, held during country workshops in Brazil, China, India, and Indonesia.

2. These national estimates are considerably lower than those of the FAO.

3. The National Environment Council (CONOMA) approved the proposal of a forestry law in March 2000, which was subsequently slotted to be presented to the National Congress by the Ministry of the Environment. This draft law resulted from numerous meetings attended by organizations representing an array of stakeholder groups. A Committee of the Chamber of Deputies had presented an alternative version of this proposed legislation to the Ministry of the Environment. The Chamber's version differed from CONOMA's proposal considerably. According to the chamber's proposal, for example, legal reserves in the Amazon region and the *Cerrados* region would occupy 50 percent and 20 percent of the territories, respectively, whereas in the CONOMA project they would occupy 80 percent and 35 percent, respectively. On May 17, 2000, the National Congress shelved the Chamber's bill in accordance with President Henrique Cardoso's pledge to oppose any reduction in the legally protected Amazon reserve area.

4. The Government of Brazil is interested in involving small and medium-size producers, whose activities in the forest sector would be subject to environmental impact assessments.

5. The 1970 Nobel Prize winner Norman Borlaug has said, "if we tried to produce the 1997 world cereal harvest using the prevailing 1960s technology, we would have needed 1.7 billion hectares of land, instead of the 700 million hectares currently in use today" (Muganda 2000).

6. Responses to the OED draft report at the workshop in Beijing November 5, 1999.

7. The Bank's latest ESW in Brazil, about factors that influence land use changes, is sound analytical work. So is the study of Cambodia's forest sector, which estimated government revenues lost through illegal logging.

8. The Bank-financed National Afforestation Project in China significantly raised the incomes of 12 million people.

9. Zoning and demarcation, among the most intensely political of issues, are particularly important challenges (Mahar and Ducrot 1998).

10. Ironically, some of the best diagnoses of the economic and political causes of tropical deforestation were conducted by World Bank staff working on Brazil forestry issues. And some of the most important lessons the World Bank has learned about the economics and politics of deforestation have been through project experience in Brazil. But these lessons have not led to a sustained productive dialogue with the Brazilian government on the future of the Amazon.

11. The government of Brazil indicated to the OED mission that the Bank's current strategy failed to meet Brazil's national goals of production in the forest sector.

Chapter 4

1. Plantations include single and multispecies plantings and agroforestry on all scales.

2. Workshops in Beijing, China, on November 5, 1999, and Brasilia, Brazil, on November 18, 1999.

3. Workshop participants in Brazil stressed the important role the Bank can play in helping Brazil develop criteria and indicators for sustainable forest management.

4. The final report of the World Commission on Forests and Sustainable Development, released in 1999, emphasized the need to make bold political decisions and develop new civil society institutions to improve governance and accountability for forest use.

5. This conclusion is strongly supported by the country studies of Brazil, India, and Indonesia.

Annex A

1. World Bank Sector Policy Papers of the 1990s are now considered Strategy Papers in the Bank. Operational Directives provided to the staff in the form of Operational Policies are considered the "policy" (see box 1.1 for details). Therefore, this review refers to the 1991 paper as either the "forest paper" or "the paper" and to its contents as the "forest strategy."

2. Based on comments of some NGOs on the OED report during ESSD consultations. Their support for Global Environment Facility activities, which often channel funds through NGOs, however, suggests that they may not be opposed to investments as much as to channeling funds through forest ministries and departments with questionable commitment to reform.

Annex B

1. Loans/credits beginning July 1991 were examined. Where available, Implementation Completion Reports and Performance Audit Reports were also reviewed, though they were not included in the analysis.

Annex G

1. ***Carbon sequestration*** refers to the process whereby forested areas retain a revolving but stable store of organic carbon in their biomass. Clearing, burning, or otherwise substantially altering the forest increases the net release into the atmo-

sphere of carbon-based gases that contribute to the greenhouse effect (World Bank OP 4.36). New initiatives such as the ***Clean Development Mechanism*** (CDM) are in the process of being developed. CDM can be seen as a partnership for organizing, structuring, and financing initiatives (involving north-south collaboration) to deal with global problems of climate change. CDM introduces the idea of "certification" as a means of evaluating collaborative programs.

Annex J

1. These guidelines are based on OP 4.36. Both OP 4.36 and GP 4.36 are based on *The Forest Sector: A World Bank Policy Paper* (The World Bank, 1991a).

2. "Bank" includes IDA, and "loans" include credits.

3. See OD 14.30, Aid Coordination Groups.

4. Specific coordination problems may arise when forestry is a responsibility of state governments within a federation. It may be necessary to broaden the scope of the policy dialogue to encompass participation at both state and national levels.

Annex K

1. World Bank 1991. *The Forest Sector: A World Bank Policy Paper*, was originally considered by the Executive Directors on July 18, 1991.

2. *World Bank Operational Manual:* Operational Policy 4.36 - Forestry.

3. *World Bank Operational Manual:* Good Practices 4.36 - Forestry.

4. "IFC Forestry Projects" (IFC/SecM91-119).

BIBLIOGRAPHY

Background Papers

Barnes, C. 1999. "Review of the 1991 'Forest Strategy: Background Paper on Monitoring and Evaluation.'" Operations Evaluation Department, World Bank, Washington, D.C.

Contreras-Hermosilla, A., Aaron Zazueta, Syed Arif Husain, Kavita Gandhi, Alejandra Martin, and Uma Lele. 2000. "Preliminary Review of the Implementation of the Bank's 1991 Forest Strategy in the Latin America and Caribbean Region." Operations Evaluation Department, World Bank, Washington, D.C.

Gautam, Madhur, Uma Lele, Saeed Rana, Hraiardi Kartodiharjo, Azis Khan, and Ir. Erwinsyah. 2000. "The Challenges of World Bank Involvement in Forests." Operations Evaluation Department, World Bank, Washington, D.C.

Husain, Syed Arif. 1999. "A Statistical Analysis of World Bank Lending Activities." Operations Evaluation Department, World Bank, Washington, D.C.

Husain, S.A., Saeed Rana, Aaron Zazueta, Kavita Gandhi, and Uma Lele. 2000. "Preliminary Review of the Implementation of the Bank's 1991 Forest Strategy in the Middle East and North Africa Region." Operations Evaluation Department, World Bank, Washington, D.C.

Kumar, N., and Uma Lele. 2000a. "Preliminary Review of the Implementation of the Bank's 1991 Forest Strategy and Its Implementation in the South Asia Region." Operations Evaluation Department, World Bank, Washington, D.C.

Kumar, N., Ridley Nelson, Uma Lele, B. Essama Nssah, Kavita Gandhi, and Alejandra Martin. 2000b. "Preliminary Review of the Implementation of the Bank's 1991 Forest Strategy in the Africa Region." Operations Evaluation Department, World Bank, Washington, D.C.

Kumar, Nalini, Naresh Chandra Saxena, Yoginder K. Alagh, and Kinsuk Mitra. 2000c. "Alleviating Poverty Through Forest Development." Operations Evaluation Department, World Bank, Washington, D.C.

Lele, Uma, Arnoldo Contreras-Hermosilla, and Kavita Gandhi. 2000a. "Preliminary Review of the Implementation of the Bank's 1991 Forest Strategy in the Europe and Central Asia Region." Operations Evaluation Department, World Bank, Washington, D.C.

Lele, U., Saeed Rana, and Kavita Gandhi. 2000b. "Preliminary Review of the Implementation of the Bank's 1991 Forest Strategy in the East Asia and Pacific Region." Operations Evaluation Department, World Bank, Washington, D.C.

Lele, U., Virgilio M. Viana, Adalberto Verissimo, Stephen Vosti, Karin Perkins, and Syed Arif Husain. 2000c. "Forests in the Balance: Challenges of Conservation with Development." Operations Evaluation Department, World Bank, Washington, D.C.

Lele, U., B. Essama Nssah, M. Gautam, and Saeed Rana. 2000d. "Limits of Environmental Adjustment for Long-Term Sustainable Growth." Operations Evaluation Department, World Bank, Washington, D.C.

Lele, U., Nalini Kumar, and Kavita Gandhi. 2000e. "Forest-Dependent Poor, Institutional Change, and the World Bank's Role." Operations Evaluation Department, World Bank, Washington, D.C.

Nssah, B. Essama, and James Gockowski. 1999. "Forest Sector Development in a Difficult Political Economy: An Evaluation of Cameroon's Forest Development and World Bank Assistance." Operations Evaluation Department, World Bank, Washington, D.C.

Rozelle, Scott, Jikun Huang, Syed Arif Husain, and Aaron Zazueta. 2000. "From Afforestation to Poverty Alleviation and Natural Forest Management: An Evaluation of China's Forest Development and World Bank Assistance." Operations Evaluation Department, World Bank, Washington, D.C.

Velozo, Ronnie de Camino, Olman Segura, Luis Guillermo Arias, and Isaac Perez. 2000. "Forest Strategy and the Evolution of Land Use: An Evaluation of Costa Rica's Forest Development and World Bank Assistance." Operations Evaluation Department, World Bank, Washington, D.C.

Other References

African Regional Workshop. 1998. "Report on the African Regional Workshop on the Underlying Causes of Deforestation and Forest Degradation." Accra, Ghana.

Binkley, C.S. 1999. "Forest in the Next Millennium: Challenges and Opportunities for the USDA Forest Service." Discussion Paper 99-15. Resources for the Future, Washington, D.C.

Biodiversity Action Network. 1999. *Addressing the Underlying Causes of Deforestation and Forest Degradation*. Washington, D.C.

Boscolo, Marco, and Jeffrey R. Vincent. 1998. "Promoting Better Logging Practices in Tropical Forests: A Simulation Analysis of Alternative Regulations." Development Research Group, World Bank, Washington, D.C.

Bowles, I., R. Rice, R. Mittermeier, and A. da Fonseca. 1998. "Logging and Tropical Forest Conservation." *Science* 280:1899–1900.

Brown, Gardner M., Jr., and Jason F. Shogren. 1998. "Economics of the Endangered Species Act." *The Journal of Economic Perspectives* 12(3): 3–21.

Calder, I. R. 1998. *Water-Resource and Land-Use Issues*. SWIM Paper No. 3. Colombo: International Water Management Institute.

Campbell, J. Gabriel, and Alejandra Martin. 2000. "Financing the Global Benefits of Forests: The Bank's GEF Portfolio and the 1991 Forest Strategy." World Bank Report Number 20627. Washington, D.C.

Cattaneo, Andrea. 1999. "Technology, Migration, and the Last Frontier: Options for Slowing Deforestation in the Brazilian Amazon." Environmental Research Service, USDA. a.cattaneo@cgair.org

Carter, Jane. 1999. "Recent Experience in Collaborative Forest Management Approaches." *Intercooperation*.

Chomitz, K., and K. Kumari. 1998. "The Domestic Benefits of Tropical Forests: A Critical Review." *The World Bank Research Observer* 13(1): 13–35.

Collier, Paul, Patrick Guillaumont, Sylviane Guillaumont, and Jan Willem Gunning. 1997. "Redesigning Conditionality." *World Development* 25(9): 1399–407.

Contreras-Hermosilla, Arnoldo. 1997. "The Cut and Run Course of Corruption in the Forestry Sector." *Journal of Forestry* 95(12): 33–36.

Dauvergne, Peter. 1997. *Shadows in the Forest: Japan and the Politics of Timber in Southeast Asia*. Cambridge, MA: MIT Press.

Delgado, C., M. Rosegrant, H. Steinfeld, S. Ehui, and C. Coutbois. 1998. *Livestock to 2020: The Next Food Revolution*. Washington, D.C.: International Food Policy Research Institute.

Dollar, David, and Jakob Svensson. 1998. *What Explains the Success or Failure of Structural Adjustment Programs?* World Bank Policy Research Working Paper, Development Research Group, Macroeconomics and Growth. Washington, D.C.

FAO (Food and Agriculture Organization). 1999a. *Sustainable Forest Management Issue Paper*. Rome.

———. 1999b. *Status and Progress in the Implementation of National Forest Programmes: Outcome of the FAO World-Wide Survey*. Rome.

———. 1999c. *State of the World's Forests*. Rome.

FAO and World Bank. 1999. "Technical Consultation on Management of the Forest Estate: Issues and Opportunities for International Action by the World Bank and FAO." Proceedings, 28–29 April 1999. Rome.

Forestry Advisory Group. 1999. *Summary Report of the 27th Meeting of the Forestry Advisory Group*. Rome.

GEF (Global Environment Facility). 1996a. *Operational Strategy*. Washington, D.C.

———. 1996b. *The GEF Project Cycle*. Washington, D.C.

———. 1996c. *Incremental Costs*. Washington, D.C.

———. 1996d. *A Framework of GEF Activities Concerning Land Degradation*. Washington, D.C.

Gregersen, Hans. 1998. "Evolving Forest Paradigms: Implications for World Bank Policy and Action." College of Natural Resources, University of Minnesota. Photocopy.

Hyde, W. 1999. "Patterns of Forest Development." Lecture given at IIED, London, April 1999.

Hyde, W., and R. Sedjo. 1992. "Managing Tropical Forests: Reflections on the Rent Distribution Discussion." *Land Economics* 68(3): 343–50.

Imazon. 1999. *Hitting the Target: Timber Consumption in the Brazilian Domestic Market and Promotion of Forest Certification*. São Paulo: FOE-Programa Amazonia; Belem, PA: Imazon; Piracicaba, SP: Imaflora.

Innes, Robert, Stephen Polasky, and John Tschirhart. 1998. "Takings, Compensation and Endangered Species Protection on Private Lands." *The Journal of Economic Perspectives* 12(3): 35–52.

Jayarajah, Carl, and William Branson. 1995. *Structural and Sectoral Adjustment: World Bank Experience, 1980-92*. Washington, D.C.: Operations Evaluation Department, World Bank.

Kaimowitz, David, and Arild Angelsen. 1997. "A Guide to Economic Models of Tropical Deforestation." Center for International Forestry Research (CIFOR), Jakarta, Indonesia.

———. 1999. "The World Bank and Non-Forest Sector Policies that Affect Forests." Center for International Forestry Research (CIFOR), Bogor, Indonesia.

Kaimowitz, David. Undated. "Protected Areas and Tropical Logging from a Political Economy Perspective." CIFOR, Indonesia. Photocopy.

Kishor, Nalin, and Luis Constantino. 1993. *Forest Management and Competing Land Uses: An Economic Analysis for Costa Rica.* Washington, D.C.: World Bank.

Kolk, Ans. 1998. "From Conflict to Cooperation: International Policies to Protect the Brazilian Amazon." *World Development* (U.K.) 26: 1481–93.

Larson, Bruce A., and Daniel W. Bromley. 1990. "Property Rights, Externalities and Resources." *Journal of Development Economics* 33:235–62.

Lele, Uma, Kinsuk Mitra, and O.N. Kaul. 1994. *Environment, Development and Poverty: A Report of the International Workshop on India's Forest Management and Ecological Revival.* Center for International Forestry Research (CIFOR), Indonesia.

London Environmental Economics Centre (LEEC). 1993. *The Economic Linkages Between the International Trade in Tropical Timber and the Sustainable Management of Tropical Forests.* London.

Lopez, Ramon. 1998. *The Tragedy of the Commons in Côte d'Ivoire Agriculture: Empirical Evidence and Implications for Evaluation Trade Policies.* Washington, D.C.: World Bank.

Lopez, Ramon, and Mario Niklitschek. 1991. "Dual Economic Growth in Poor Tropical Areas." *Journal of Development Economics* 36: 189–211.

Mahar, Dennis J., and Cecile E. H. Ducrot. 1998. *Land-Use Zoning on Tropical Frontiers: Emerging Lessons From Brazilian Amazon.* WBI Case Studies, World Bank Institute. Washington, D.C.: World Bank.

Metrick, Andrew, and Martin L. Weitzman. 1998. "Conflicts and Choices in Biodiversity Preservation." *The Journal of Economic Perspectives* 12(3): 21–34.

Muganda, Clay. 2000. "One Man's Relentless Effort to Provide Bread for the Hungry." *Kenya Daily Nation,* June 8, 2000.

Munasinghe, Mohan. 1993. "The Economist Approach to Sustainable Development." *Finance and Development* 30(4).

North, Douglas. 1990. *Institutions, Institutional Change and Economic Performance.* Cambridge, U.K.: Cambridge University Press.

Pagiola, Stefano, John Kellenberg, Lars Vidaeus, and Jitendra Srivastava. 1997. *Mainstreaming Biodiversity in Agricultural Development.* Washington, D.C.: World Bank.

Palo, Matti, and Jussi Uusivuori, eds. 1999. *World Forests: Society and Environment.* Dordrecht/London/Boston: Kluwer Academic.

Panayotou, T. 1995. "Environmental Degradation at Different Stages of Economic Development." In: I. Ahmed and J.A. Doeleman, eds., *Beyond Rio: The Environmental Crisis and Sustainable Livelihoods in The Third World.* London: Macmillan.

Pearce, David, Francis Putz, and Jerome K. Vanclay. 1999. *A Sustainable Forest Future?* London: Natural Resources International and U.K. Department for International Development.

Persson, Reidar. 1998. *From Industrial Forestry to Natural Resources Management: Lessons Learnt in Forestry Assistance.* Stockholm; Swedish International Development Cooperation Agency.

Picciotto, Robert. 1992. *Participatory Development: Myths and Dilemmas.* World Bank Working Paper No. 930. Washington, D.C.

Picciotto, Robert, and Eduardo Wiesner. 1998. *Evaluation and Development: The Institutional Dimension.* Washington, D.C.: World Bank.

Poore, Duncan, with Jurgen Blaser, Eberhard F. Bruenig, Peter Burgess, Bruce Cabarle, David Cassells, Jim Douglas, Don Gilmour, Pat Hardcastle, Gary Hartshorn, David Kaimowitz, Nalin Kishor, Alf Leslie, John Palmer, Francis Putz, M.N. Salleh, Nigel Sizer, Timothy Synott, Frank Wadsworth, and Tim Whitmore. 1998. "No Forest Without Management. Sustaining Forest Ecosystems Under Conditions of Uncertainty." Submitted to *Science.*

Quality Assurance Group (QAG), World Bank. 1998. *Quality of ESW in FY98.* Washington, D.C.

Rice, Richard, Cheri Sugal, and Ian Bowles. 1997. *Sustainable Forest Management: A Review of the Current Conventional Wisdom.* Washington, D.C.: Conservation International.

Rich, Bruce. 1993. *Mortgaging the Earth: The World Bank, Environmental Impoverishment, and the Crisis of Development.* Boston, MA: Beacon.

Ross, Michael. 1996. "Conditionality and Logging in the Tropics." In Robert O. Keohane and Marc A. Levy, eds., *Institutions for Environmental Aid: Pitfalls and Promise.* Cambridge, MA: MIT Press.

Ruitenbeek, J., and C. Cartier. 1998. *Rational Exploitation: Economic Criteria & Indicators for Sustainable Management of Tropical Forests.* Indonesia: CIFOR.

San, Nu Nu, H. Lofgren, S. Robinson, and S.A. Vosti. 1998. *Macroeconomic Policy, Labor Migration, and Deforestation in Sumatra: Progress Report*. Washington, D.C.: International Food Policy Research Institute.

Sandler, Todd. 1997. *Global Challenges: An Approach to Environmental, Political and Economic Problems.* Cambridge, U.K.: Cambridge University Press.

Schneider, Robert. 1992a. *Brazil: An Analysis of Environmental Problems in the Amazon,* Vol. 1: *Main Report*. Report No. 9104-BR. Washington, D.C.: World Bank.

———. 1992b. *Brazil: An Analysis of Environmental Problems in the Amazon,* Vol. 2: *Annexes*. Report No. 9104-BR, Washington, D.C.: World Bank.

Schneider, Robert, Adelberto Verissimo, and Virgilio Viana. 1998. "Logging and Tropical Forest Conservation." Submitted to *Science*.

Schneider, Robert, E. Arima, P. Barreto, and A. Verissimo. 2000. "Sustainable Forestry and the Changing Economics of Land: The Implications for Public Policy in the Legal Amazon." Economic and Sector Work, ESSD, World Bank, Washington, D.C. Photocopy.

Seymour, Frances, and N. Dubash. 2000. *The Right Conditions: The World Bank, Structural Adjustment and Policy Reform*. Washington, D.C.: World Resources Institute.

Shepherd, Gill, Mike Arnold, and Steve Bass. 1999. *Forests and Sustainable Livelihoods*. Washington, D.C.: World Bank.

Shogren, J., and J. Tschirhart. 1999. "The Endangered Species Act at Twenty-Five." *Choices*, 3rd quarter.

Simula, Markku, and Indufor Oy. 1999. *Certification of Forest Management and Labeling of Forest Products*. Washington, D.C.: World Bank.

Spears, John. 1998. "Sustainable Forest Management: An Evolving Goal." A position paper prepared for UNDP's Global Programme. Washington, D.C.

Spilsbury, Michael J., and David Kaimowitz. Undated. "The Influence of Policy Research and Publications on Conventional Wisdom and Policies Affecting Forests." Center for International Forestry Research, Bogor, Indonesia.

Stern, David I., Michael S. Common, and Edward B. Barbier. 1996. "Economic Growth and Environmental Degradation: The Environmental Kuznets Curve and Sustainable Development." *World Development* 24(7): 1151–60.

Struhsaker, T. T. 1998. "A Biologist's Perspective on the Role of Sustainable Harvest in Conservation." *Conservation Biology* 12(4): 930–32.

Thiele, Rainer, and Manfred Wiebelt. "Policies to Reduce Tropical Deforestation and Degradation: A Computable General Equilibrium Analysis for Cameroon." *Quarterly Journal of International Agriculture* 33(2): 162–78.

Tomich, Thomas P., David E. Thomas, Y. Kusumanto, and Meine van Noordwijk. 1998. "Policy Research for Sustainable Upland Systems in Southeast Asia." International Center for Research in Agroforestry, Kenya.

Tomich, Thomas P., Meine van Noordwijk, Stephen A. Vosti, and Julie Witcover. 1998. "Agricultural Development with Rainforest Conservation: Methods for Seeking Best Bet Alternatives to Slash-and-Burn, with Applications to Brazil and Indonesia." *Agricultural Economics* 19(1998):159–74.

United Nations. 1999a. *Human Development Report*. New York.

———. 1999b. *Report of the Secretary-General*. Geneva.

Vosti, S.A., J. Witcover, S. Oliveira, and M.D. Faminow. 1997. "Policy Issues in Agroforestry: Technology Adoption and Regional Integration in the Western Brazilian Amazon." *Agroforestry Systems* 38(1-3):195–222.

Whiteman, A., C. Brown, and G. Bull. 1999. *Forest Product Market Developments: The Outlook for Forest Product Markets to 2010 and the Implications for Improving Management of the Global Forest Estate*. Rome: FAO, Forest Policy and Planning Division.

Witcover, J., and S.A. Vosti. 1995. "Workshop on Non-Timber Tree Product (NTTP) Market Research." Environment and Production Technology Division, (EPTD) Workshop Summary Paper No. 3, International Food Policy Research Institute, Washington, D.C.

World Bank. 2000a. "Implementation of the 1991 Forest Strategy in IFC's Projects." OEG. Washington, D.C.

———. 2000b. *Financing the Global Benefits of Forests: The Bank's GEF Portfolio and the 1991 Forest Strategy*. OED. Washington, D.C.

———. 1998a. *Summary Report: Study of GEF Projects Lessons*. Washington, D.C.

———. 1998b. *Source and Sinks of Greenhouse Gases.* GEF. Washington, D.C.

———. 1998c. *Financial Sector Reform: A Review of World Bank Assistance.* OED Report No. 17454. Washington, D.C.

———. 1998d. "Approach Paper: Forests and the World Bank, An Operations Evaluation Department Study." Washington, D.C.

———. 1998e. "Forest Policy Implementation Review and Strategy: Initiating Memorandum," ESSD. Washington, D.C.

———. 1998f. "Sustaining Tropical Forests: Can We Do It, Is It Worth Doing?" Tropical Forests Expert Discussion Meeting, Meeting Report, Part I. Forests Team, Natural Resources Program, Environment Department. Washington, D.C.

———. 1998g "Sustaining Tropical Forests: Can We Do It, Is It Worth Doing?" Tropical Forests Expert Discussion Meeting, Meeting Report, Part II. Forests Team, Natural Resources Program, Environment Department. Washington, D.C.

———. 1998h. *Environment Matters. Annual Review.* Washington, D.C.

———. 1998i. "East Asia Natural Resource Management Strategy." Washington, D.C. Photocopy.

———. 1998j. *Assessing Development Effectiveness: Evaluation in the World Bank and the International Finance Corporation.* OED. Washington, D.C.

———. 1998k. "Biodiversity in World Bank Projects: A Portfolio Review." World Bank Environment Department Working Paper No. 59. Washington, D.C.

———. 1997a. *What Drives Deforestation in the Brazilian Amazon? Evidence from Satellite and Socioeconomic Data Environment Working Paper.* Report Number WPS1772. Washington, D.C.

———. 1997b. "Portfolio Improvement Program: Natural Resource Management Portfolio Review." Environment Department. Washington, D.C.

———. 1996a. *Forestry Portfolio Review.* AGR/QAG. Washington, D.C.

———. 1996b. *Resettlement and Development: The Bank-Wide Review of Projects Involving Involuntary Resettlement, 1986–1993.* Environment Department. Washington, D.C.

———. 1996c. "The Inspection Panel Report: August 1, 1994 to July 31, 1996." Washington, D.C.

———. 1996d. *The Evaluation of Economic and Sector Work: A Review.* Operations Evaluation Department. Washington, D.C.

———. 1995a. *Working with NGOs: A Practical Guide to Operational Collaboration Between the World Bank and NonGovernmental Organizations.* Operations Policy Department. Washington, D.C.

———. 1995b. *Performance Indicators in Bank-Financed Agricultural Projects.* Washington, D.C.

———. 1995c. *Environmental and Economic Issues in Forestry.* Washington, D.C.

———. 1994a. *Review of Implementation of the Forest Sector Policy.* Agricultural Department. Washington, D.C.

———. 1994b. *Conditional Lending Experience in World Bank-Financed Forestry Projects.* OED Report No. 13820. Washington, D.C.

———. 1994c. *Global Environmental Facility: Independent Evaluation of the Pilot Phase.* Washington, D.C.

———. 1993a. "The World Bank Operational Manual, Operational Policies: Forestry." Report Number OP4.36. Washington, D.C.

———. 1993b. "The World Bank Operational Manual, Operational Policies: Natural Habitats." Report Number OP4.04. Washington, D.C.

———. 1993c. "The World Bank Operational Manual, Operational Policies: Environmental Action Plans." Report Number OP4.02. Washington, D.C.

———. 1993d. "The World Bank Operational Manual, Bank Procedures: Natural Habitats." Report Number BP4.04. Washington, D.C.

———. 1993e. "The World Bank Operational Manual, Operational Directives: Environmental Policies." Report Number OD4.00. Washington, D.C.

———. 1993f. "The World Bank Operational Manual, Operational Directives: Environmental Assessment." Report Number OD4.01. Washington, D.C.

———. 1993g. "The World Bank Operational Manual, Operational Directives: Poverty Reduction." Report Number OD4.15. Washington, D.C.

———. 1993h. "The World Bank Operational Manual, Operational Directives: Indigenous People." Report Number OD4.20. Washington, D.C.

———. 1993i. "The World Bank Operational Manual, Operational Directives: Involuntary Resettlement." Report Number OD4.30. Washington, D.C.

———. 1992a. *Guidelines for Monitoring & Evaluation of GEF Biodiversity Projects.* Environment Department. Washington, D.C.

———. 1992b. *World Bank Approaches to the Environment in Brazil: A Review of Selected Projects.* OED Report No. 10039. Washington, D.C.

———. 1992c. *Brazil: An Analysis of Environmental Problems in the Amazon.* OED Report No. 9104-BR. Washington, D.C.

———. 1991a. *The Forest Sector: A World Bank Policy Paper.* Washington, D.C.

———. 1991b. *Forestry: The World Bank's Experience.* Operations Evaluation Department. Washington, D.C.

———. Undated. "Investing in Destruction—The World Bank and Biodiversity." Washington, D.C.

World Bank, ESSD Network. Undated. "Europe and Central Asia: Forest Policy Sector Notes." Washington, D.C. Photocopy.

World Rainforest Movement, Forest Peoples Programme and Environmental Defense Fund. 1999. *Briefing Paper: The World Bank's Forest Policy Implementation Review and Strategy Development.* www.wrm.org.uy/english/tropical_forests/wb.html

World Wide Fund for Nature (WWF). 1999. *Conversion of Paper Parks to Effective Management—Developing Target.*

WWF and International Union for the Conservation of Nature (IUCN). 1999. *Management Effectiveness in Forest Protected Areas.* Gland.

FORESTRY EVALUATION COUNTRY CASE STUDY SERIES

Brazil: Forests in the Balance: Challenges of Conservation with Development
Cameroon: Forest Sector Development in a Difficult Political Economy
China: From Afforestation to Poverty Alleviation and Natural Forest Management
Costa Rica: Forest Strategy and the Evolution of Land Use
India: Alleviating Poverty through Forest Development
Indonesia: The Challenges of World Bank Involvement in Forests

OPERATIONS EVALUATION DEPARTMENT PUBLICATIONS

The Operations Evaluation Department (OED), an independent evaluation unit reporting to the World Bank's Executive Directors, rates the development impact and performance of all the Bank's completed lending operations. Results and recommendations are reported to the Executive Directors and fed back into the design and implementation of new policies and projects. In addition to the individual operations and country assistance programs, OED evaluates the Bank's policies and processes.

Summaries of studies and the full text of the Précis and Lessons & Practices can be read on the Internet at **http://www.worldbank.org/html/oed/index.htm**

How To Order OED Publications
Operations evaluation studies, World Bank discussion papers, and all other documents are available from the World Bank InfoShop.

Documents listed with a stock number and price code may be obtained through the World Bank's mail order service or from its InfoShop in downtown Washington, D.C. For information on all other documents, contact the World Bank InfoShop.

For more information about this study or OED's other evaluation work, please contact Elizabeth Campbell-Pagé or the OED Help Desk.

Operations Evaluation Department
Partnerships & Knowledge Programs (OEDPK)
E-mail: ecampbellpage@worldbank.org
E-mail: OED Help Desk@worldbank.org
Telephone: (202) 473-4497
Facsimile: (202) 522-3200

Ordering World Bank Publications
Customers in the United States and in territories not served by any of the Bank's publication distributors may send publication orders to:

The World Bank
P.O. Box 960
Herndon, VA 20172-0960
Fax: (703) 661-1501
Telephone: (703) 661-1580

The address for the World Bank publication database on the Internet is:

http://www.worldbank.org/publications

The World Bank InfoShop serves walk-in customers only. The InfoShop is located at:

701 18th Street, NW
Washington, DC 20433, USA
E-mail: pic@worldbank.org
Fax number: (202) 522-1500
Telephone number: (202) 458-5454

All other customers may place their orders through their local distributors.

Ordering by e-mail
If you have an established account with the World Bank, you may transmit your order by electronic mail on the Internet to: **books@worldbank.org.** Please include your account number, billing and shipping addresses, the title and order number, quantity, and unit price for each item.